THE NIGHT
the Sirens Blew

by

Allen W. Taylor

TORNADO PRESS

MINNEAPOLIS

The Night The Sirens Blew

First published by:

Tornado Press

12144 Undercliff Street NW

Coon Rapids, MN 55433

763-422-3878

Photos are used with permission, courtesy of
Minnesota Historical Society
Fridley Historical Society

Library of Congress Cataloging-in-Publication
2008905039

Printed in United States of America

1st Printing Dickinson Press
June 2008 - ISBN 978-1-60585-712-1

2nd Printing CreateSpace
February 2011 - ISBN 978-1-45656-871-9

The clouds poured out water;

the skies gave forth thunder;

thy arrows flashed on every side.

The crash of thy thunder was

in the whirlwind;

thy lightnings lighted up the world;

the earth trembled and shook.

Psalm 77: 17-18

Acknowledgments

There are no words to express my gratitude to the many gifted professionals who worked with me on this book. So many people gave of themselves for the pursuit of creating this book. Numerous special individuals warmly accepted me into their lives and professions and for that I am grateful. Working with the following people was the event of a lifetime.

Charlie Boone, WCCO Radio Personality

Dick Chapman, WCCO Radio Personality (retired)

Rob Brown, WCCO Radio Executive (retired)

Robert Christenson, President-Fridley Historical Society, Fridley History Center Board

Ellen Miller M.A., Minnesota Historical Society, Exhibit Developer

Todd Krause, National Weather Service-Chanhassen

Daphne LaDue-Phd. Candidate-University of Oklahoma/Norman

Dr. Harold Brooks, National Severe Storms Laboratory/Norman

Carol Allen, Independent School District 11 Anoka Hennepin

Kathy Serpinas, Independent School District 16
Spring Lake Park

Jerry Stegeman, Private Citizen Archive Collection

Donald Hodges, Former Mayor of Mounds View,
Private Archive Collection

Jodee Kulp, Better Ending New Beginnings,
Typography/Book Design

I also want to thank

the hundreds of kind,

generous people whose stories

and recollections

make this book so

informative and fascinating!

Dedication

I dedicate this book to Almighty GOD, my wife Kathy, my sons Jacob and Noah, and to the individuals, families, and friends who suffered devastating losses from the tornadoes of May 6, 1965.

A portion of the proceeds

from the sale of this book

will be donated to the

Fridley Historical Society

and the Minnesota Historical

Society for their

tireless dedication

to the preservation

of Minnesota History.

Introduction

It is said that the events in our childhood shape our world and our perceptions. The origins of this book go back to the evening of May 6th, 1965. I was five years old, without the awe and inspiration I now have for storms. The vivid, surreal colored lightning and the once-in-a-generation power associated with these storms left an impact on me and all the people who lived through those storms. Those storms changed the way we think about storms and that we are vulnerable. The two main reasons, I discovered in doing this book, that there were so few fatalities was we lived in an age of the Cold War and if the sirens blew, you sought shelter immediately. Civil Defense was still in place and everyone knew where to go if the sirens blew. The other reason was due to the excellent coverage of WCCO. Joe Strub Jr., a fantastic meteorologist

at the Weather Bureau, was known to have said that WCCO Radio saved two to three thousand lives that night. As you read about the damage and the personal accounts of the survivors, you understand why.

The book began about a year ago and the greatest joy has been meeting and talking with people who lived through those tornadoes. Stories of heroism, sheer terror, and the drive to start over was incredibly inspirational. Those tornadoes wiped out everything! In today's dollars, damage was estimated at $1.2 billion dollars, making this swarm of tornadoes one of the costliest outbreaks in U.S. history! There were three main storm surges which spawned 24 tornadoes in one evening. It is my hope that you feel the emotions of the people who survived that night.

Some of the information might be slightly embellished due to the excitement felt that evening but I found almost all of the stories to match and be highly factual. Many thousands of people were affected and left with scars that still remain to this day.

The cobweb dampness of the basement added to the raw energy of the storm. The wild multi-colored lightning and ear-splitting thunder accentuated the sound of the hail hitting

the windows and sides of the house. The calm reduction of sound to an eerie quiet gave way to a roar. Tremendous pressure along with the thunderous noise of a tornado shattered the calm that existed before. To what seemed an eternity, the storm was over. Surveying the unbelievable damage and the numbness of losing everything was felt over and over again that night.

This book is split into two main parts. The first is the highly descriptive, emotional personal accounts. Many of the stories sent chills down my spine. I can't begin to imagine the emotions felt that day. Emotions that lingered long after the tornadoes departed. The second is the dynamics of the storm. These dynamics include Weather Bureau records and other historical archive records. These records present valuable insight into the events leading to the formation of these legendary storms.

It is my hope that you enjoy the most researched, documented book ever written on the worst tornado outbreak ever to hit the Twin Cities

Allen W. Taylor

May 2008

Charlie Boone, Famous WCCO Radio personality

(stayed on duty to help inform public of tornadoes)

That was an unbelievable night! I was on duty at the end of the Franklin Hobbs & Howard Viken shows when "Chappy" (Dick Chapman) came on. I wanted to leave right away to cover some news developing west of the Twin Cities due to the violent storms being reported. But, conditions here deteriorated so rapidly here that I decided to stay and help man the phones with all the calls coming in. Our production person, Rob Brown also helped with the incredible number of calls coming in. At first, we didn't believe that there were that many tornadoes happening and that people were seeing tornadoes that weren't there. As the night progressed and all the calls did correlate with the actual sightings and radar, we knew that the tornadoes were for real!

I'll never forget when a young man called in from Fridley and his name was Robert Clarke. Later that evening, he was killed by another tornado that had come through that area. When you're on the air for a long time, you become completely absorbed by the event. We began

reporting on the tornadoes at about 5:30 p.m. until around 11:00 p.m. So much information was coming in and I had never heard of so many tornadoes happening at the same time! I felt way over my head but "Chappy" was there and he was the real hero. He was the true newsman. His calm demeanor and ability to stay calm and focus on the story really helped all of us in the newsroom. You have to remember that our audience in those days was tremendous! Everyone listened to 'CCO and we really had an impact on our audience.

The year 1965 was a terrible weather year. We had a record snow winter followed by unbelievable flooding and then, the terrible tornadoes. In fact, our reporting of those events won WCCO three prestigious news awards that year including the Peabody Award. I represented 'CCO in New York City to accept the award and sat next to our First Lady, Mrs. Johnson. It was a humbling experience sitting at the head table with all the CBS greats: Walter Cronkite, Mike Wallace, etc.

The evening of May 6th, 1965 was a night I will never forget!

Dick Chapman, Famous WCCO radio personality

(main broadcaster the night of the tornadoes)

That night began for me as I was on the air shortly after 6 p.m. The first funnel was sighted near Norwood-Young America, just west of the Twin Cities. After promptly broadcasting the first warning it became painfully evident it was not one storm. From that point on, I never left the microphone until almost 1 a.m. the next morning.

I had the privilege (duty actually) to head storm coverage that saved countless lives and captured the top three national awards in broadcast journalism-The Sigma Delta Chi/Professional Journalist Society, The Alfred I. Dupont, and the prestigious Peabody Award. No one has ever swept all three. But, the gift that I treasure from that time is a piece of brown paper signed by nine cub scouts from the Fridley area thanking me for saving their lives! Joe Strub, assistant Weather Bureau chief, told me that our coverage of those tornadoes saved 2000-3000 lives. WCCO Radio was prepared because of the horrendous blizzards of the 1965 winter and the ensuing floods that wracked the state. We had installed additional phone service and set up

emergency procedures, never realizing how important that would be until the funnels started boiling down. I counted nine separate tornado touchdowns in the Twin Cities that night. The "heat dome" prevented many of those from actually hitting the Minneapolis city proper.

But the real story of the evening was the performance of the WCCO Radio listeners. They became our eyes in the field. They tracked the funnels with absolutely amazing accuracy! We could put them on the air live with all our phone lines. On at least one occasion, I remember three separate "spotters" using a triangular method to pinpoint a massive funnel's location and describe its movement block by block. It came so close to one of these citizens he had to hurriedly disconnect and race to his basement. I had a caller tell me that our reports actually coincided with the systematic ripping apart of his house, right on the air!

These spotters/listeners were sensational-not a crackpot among them. The dire conditions of that night brought out civic responsibility unheard of. I used a simple city map to coordinate the reports. Charlie Boone and Rob Brown joined me later to help screen the callers. And on and on it went: heavy funnel damage first in the Lake Minnetonka area-Navarre was leveled, Spring Lake Park, Excelsior, and Wayzata hit hard. It was difficult to believe that there was more than one funnel on the ground at any given time.

John Walker, one of our reporters, called in with a report of a tornado actually hitting the Wayzata Boat Works where he was at. He had to disconnect while the tornado had blown glass all over and overturned boats right at his location. He hid under a big desk and called a minute or two after the tornado had moved on! It was incredible! I had a caller state that he was driving a VW by Black & Seton Lakes watching a tornado, unbeknownst to him, there was another tornado coming up from behind him which threw him and his car into Lake Minnetonka. He wound up on shore and watched his car submerge under the surface with the headlights still shining! He was in a daze and didn't know how he made it to shore because he didn't know how to swim! It was just hard to understand that much power in one storm system.

Afterwards, I had a conversation with Joe Strub from the Weather Bureau. He was an outstanding guy who really knew a lot about storms. At the time, he was working on tornado research

and found that these storms would "pulse". In other words, the storms would build-drop-build-drop. This would help in future tornado warning because you could time the cycle and issue the advance warning. He explained that this storm was so massive that it literally "chewed" up debris and flung it all over the U.S. They found debris as far away as Pennsylvania! Joe was also a pioneer in lobbying for safer trailer park tornado prevention measures. He wanted "cement tie-downs" and reinforced storm shelters in every trailer park. Finally, after some years, the legislature enacted some of these measures that actually saved lives in subsequent tornado outbreaks in Minnesota.

About halfway through that evening, there were reports of funnels being sighted over Southdale. I was so busy that I realized my family was in close proximity to that tornado. There was nothing I could do and I felt helpless in the situation. I had no way of knowing if my family was OK or not. Fortunately, my family had sought shelter in the basement and no damage occurred at my home. In Mounds View, I saw homes that no longer existed and tons of debris in the basement. There were no safe corners to hide in those homes from the tornadoes! At the time, it was recommended that you took shelter in the SW corner of the basement. Seeing all the debris, it would only be safe under heavy furniture or stairs. The wind in those storms was incredible! I saw many telephone poles with pine needles driven into them. In the helicopter, you could see where the tornadoes had actually "clipped" the tree tops like a lawnmower.

It was an evening I will never forget!

WCCO staff pictured here kept the public advised for 6 1/2 hours the evening of May 6th, 1965.

Pictured (l to r): Dick Chapman, calling out another list of communities in the path of the tornadoes;Rob Brown, processing calls from the public; Bob Tibbitts, in contact with John Walker at Lake Minnetonka; Charlie Boone, ready with another "live sighting" from a listener.

Bob Brown,
Promotion Director WCCO Radio-1965

(helped man the phones with Charlie Boone and Dick Chapman during the tornado outbreak)

That evening brings a flood of memories for me. An interesting side note to that night was that Ray Christianson and I completed a demo tape for tornado warnings on April 24, about 2 weeks prior to the actual tornado outbreak of May 6, 1965. When we did it, people would comment, "Why would you want to do a tornado warning tape? We never get tornadoes around here!" Sure enough, 2 weeks later we had the May 6th tornadoes. There were so many lucky events that happened prior to the tornadoes. We had just gone through the worst winter and spring flooding of all time and we still had studio 2 intact. Several times, people asked if we should dismantle the phones, wall maps, etc. Something told me inside not to do it and when the tornadoes hit, we were ready for it. For the blaring warning tone, I used a heavy, chrome two-piece ashtray from the 30's. We had a Degausser demagnetizing tool that we placed on the ashtray with an on/off switch. When you turned it "on", the magnet would vibrate (magnetic flux) the top of the ashtray causing that awful, attention-getting noise! With the demo and that warning tone, it was called the "red disk". Now, I was leaving my drills at Fort Snelling and was still wearing my army uniform when I looked down the Minnesota River valley. It was hot, humid, and the sky was green. I turned on the radio and heard the "red disk" being played and only a handful of people knew the story behind that warning. I looked down the valley and saw "fingers" going up and down and knew we were in trouble. I hurried to the station in downtown Minneapolis to help out. Charlie and I were answering the phones while Dick was telling the public where the tornadoes were. Dick did the most impressive job that I had ever seen. He was and is a "newsman". Afterwards, he told me that we took a tremendous journalistic risk by putting the public on the air. Honest to goodness, I know of only 1 or 2 kids who called in with prank calls and the rest were legit! You have to remember, back in those days we controlled over 68% of the market share and by the time the evening ended, other radio stations and even the Weather Bureau were listening to what we had to say. A good friend of mine, Joe Strub told me that 'CCO saved 2000 lives that night. I believe it.

The news was unfolding in front of our eyes and we just went with it! I think with all the bad weather we had that year, we were ready and the public was ready for instant news. We just went through the St. Patrick's Day Blizzard and the floods. In the blizzard, snowmobiles were used to transport blood to the hospitals. It was unbelievable! At first, we didn't believe that there were that many tornadoes on the ground but as the calls kept coming in, we realized that it was true and all we could do was report the news and hopefully save lives in front of the storms heading in that direction. After the storms, the Weather Bureau hooked in a "tie-line" between WCCO and their offices. So, when severe weather hit, all we had to do (and vice versa) was pickup the phone and instant reports right then and there! As the night progressed we kept hearing of tornadoes hitting Fridley with the last one striking at about 10 p.m. I'll never forget the two calls I got that night. First, Robert Clarke called in from Fridley and reported on that tornado hitting the Fridley Junior High School, later on that night he was killed when the next tornado hit. Second, I was leaving at about 1 a.m. when our receptionist told me that a gentleman from Fridley wanted to talk to someone who was in the call center that night. So, I picked up the phone

and this gentleman from Fridley thanked me for saving him and his family from the devastating tornadoes that went through Fridley that night. He went on to explain what happened. The first tornado came through and they heard the warnings over 'CCO and they went to a sandy, crawlspace under the garage. After that first tornado went through, they went outside to look at the damage and only some window/shingle damage occurred. We informed the public of a second tornado heading for that area, so they went back to the crawlspace. This time they lost their porch. The third warning came out and they made it to their shelter and this time, lost their entire house. He explained that the timely, informative reports had truly saved his family from what would have been sure death. You could tell that this guy was very emotional and it really struck me deep. I knew it was going to be a long evening when the first casualty was a farmer from Norwood named Raymond Perbix, this happened around 5 p.m.

Joe Strub, God rest his soul, was a great guy and a good friend. He told me after the tornadoes that Phil Kenworthy and he had worked a project named "Gen-Con". This was short for "Generation Continuum". Phil and Strub found that tornadoes in Minnesota would vent and repeat every 22 minutes. Amazingly, Strub told me that the "tops" of those storms which came over were at about 90,000 feet! There were three ripples in the storm system that covered Minnesota that day. They were located from Fergus Falls to Hill City, Waseca to Red Wing, and of course, Lake Minnetonka to Anoka.

After the tornadoes, I went to Mounds View to see the damage with Dick and was absolutely astounded by the damage. There was a house that was still standing with all types of warning signs telling people to not go in or be around it. The house didn't look that bad and we couldn't figure out why they (authorities) would have such warnings. A neighbor came over and told us that the tornado had picked up the roof and sucked all the nail-heads out of that house and set the roof back on top. The house could literally crumble at any time! On Lois Drive, an area especially hit hard, we saw what used to be a house that looked like a swimming pool! The tornado sucked everything out of the basement and the house was completely gone. That tornado hit so hard and quick, it "mowed" the house of the cinder block foundation. You know how the mortar would be the cut line, this tornado broke the blocks in half at the grass level! Furthermore, the tornado had sucked all the utilities out of the ground leading up to the house!! No wiring, no plumbing, everything was sucked up by the tornado! **It was an unbelievable night!**

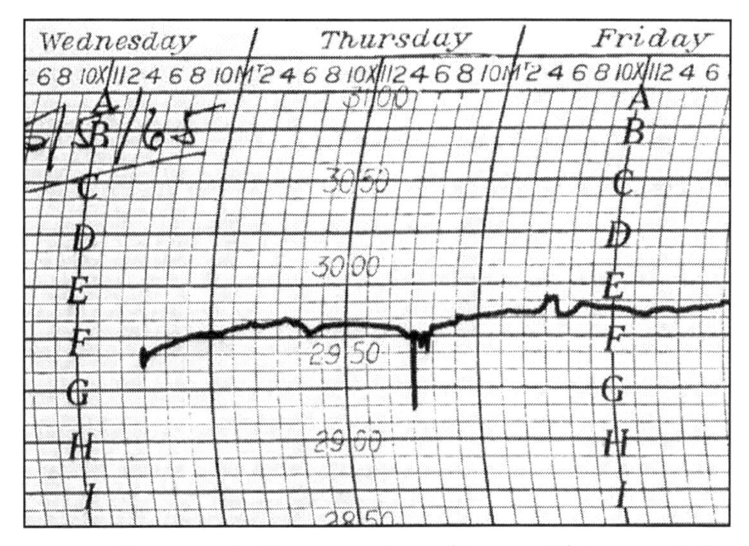

Barograph (barometer) reading at Christmas Lake in direct path of tornado. Note the momentary pressure drop from 29.50 to 29.13. The next short drops indicate the close proximity of the tornado that struck from the west.

Don & Amy Hodges, 5086 Eastwood Road, Mounds View

(Don was the mayor of Mounds View in 1965)

The tornado hit Mounds View with renewed fury after hitting Fridley. We were at a Boy Scout meeting in the basement of the Spring Lake Park Lutheran Church at the corner of Red Oak & Hillview. It was dark due to the electricity that was cut by previous tornadoes in the area. The speaker of that night was Judge Gingold, who wouldn't stop talking. Lee Erickson, one of the scoutleaders, came rushing down the stairs stating that he had heard on the car radio that a big tornado was heading our way. He had heard that tornadoes were bouncing around all over the place! We all rushed home and on the way, there was debris on the roads; lawn chairs and other stuff. When we got home, we looked from our driveway and you could see the monster tornado coming our way! It looked like a big cloud coming out of Fridley. I almost beat my family down to the basement! My wife, our 6 kids, and I huddled in the basement and you could hear the roar of the tornado! Amy and I were "storm warned" from growing up in Iowa and knew that this was bad!

The tornado first hit Mounds View on the corner of Knollwood & County Road H2 where the Messiah Lutheran Church stood. The tornado hit with such force that it cracked the heavy wooden arch beams in the ceiling. Kittleson's Tom Thumb store was across the street and that building was destroyed. A gentleman named Achterkirch, owned a small lumber business, was in the back of the Tom Thumb moving such stuff and was killed. The authorities found his body in a tree down the block from that store. That tornado tore through the old Johnson farm to Highway 10 where a young man named Clarke was killed in his pickup truck. The storm just missed our City Hall and went over the area that is now the Mounds View Square. The twister then hit Sunrise Methodist Church and destroyed the parsonage on the corner of County Road I & Long Lake Road.

Usually tornadoes weaken at the end of their destructive track but this one redoubled in fury as it hit the Lois Drive area. Ray Young was taking a nap on his davenport and woke up in the swamp that was a few blocks away from his home on Lois Drive. That entire area of Lois Drive, Hillview Road, and Knoll Drive was completely obliterated. This area is now known as the Colonial Village Park area. All you could see was small pieces of debris. The authorities found a little baby buried in the rubble. When they found the little girl, they thought it was a doll. Many people were killed in this area. When I saw the devastation, it made you gasp. It was heartbreaking to see what had happened to these poor people. The houses were gone to their foundations! The looters came out right away and we had to call the National Guard in to supervise that area. You needed a pink pass to enter or leave the area. I forgot my pass on one occasion and a guardsman made me go back to City Hall to get another one, which I did. I was glad to see that they were taking their important jobs serious.

In the aftermath, Governor Karl Rolvaag was tremendous! He came out right away to see the devastation and he worked hard to make sure that we had the resources we needed for clean-up, etc. Even months after the tornado, I would get calls from Karl asking how things were going and if we needed anything. He was a wonderful man! Empty HUD were opened for the homeless and I happily distributed the house keys to those families in great need. During those days following the tornado, I doubt if I slept at all. You really ran on adrenaline!

Amy: I was just amazed at how the community came together to help each other! If you needed clothes washed, a babysitter, or cleaning houses, there were volunteers willing to do anything to help! People were so generous from other communities as well. Food, clothing, cleaning supplies, whatever was needed was given. Everyone pitched in to help! Today, things are different. Volunteers are far and few between. The next morning was rainy and sky was an icky green color. The amount of debris and people who lost their papers, titles, and other important papers was terrible. The following Memorial Day weekend, we spent that holiday sorting through boxes of papers found in the debris. We were able to distribute a lot of papers back to their original owners.

Georgia Buck, 2303 Lois Drive, Mounds View

As the day wore on, I noticed it was getting very muggy and warm. I figured that when the kids got home they'd put on their summer shorts for the first time that year. Sure enough, as soon as they got home, off went the school clothes, shoes and on went the shorts. They thought it was great fun wearing shorts on May

6th. Normally, here in Minnesota it's still slacks and sweater weather.

About 5:30 or 6:00 it started to rain. It just poured! Our radio is on almost all the time, so when they started to announce a Tornado Watch for the Twin Cities, my ears perked up. I told the kids to stay in the yard in case of a storm. Our next door neighbor works at the same trucking terminal as Duane, my husband, but he is on an earlier shift so he gets home earlier. I looked out the front door and saw him and his wife, out in the rain tying down a plastic cloth to protect the new top on their Jeep from the rain. I asked if he had seen Duane and if he was on his way home because there was a Tornado Watch and we were included in that area. Then it got really dark! I didn't want to frighten the children, but I kept hoping my husband would get home any second. In the meantime, I somehow made supper and fed the kids. About 6:15 Duane drove up the driveway and I had Pat, our oldest daughter, run out in the rain and open the garage door for him so he wouldn't get wet. He had heard about the Tornado Watches but, of course, who believes it could happen to you.

Duane and the kids kept going in and out of the house watching the sky hoping to see something happen. All of a sudden there was a funny smell in the air and the sky became green. They saw a tornado hit an airfield a mile or so away. A big piece of something fell in our backyard and Duane said to Mike, our oldest son, "Tomorrow morning we'll go out and see what that is!" He never did find out. As we listened to the radio, it seemed the storm was not over and that several funnels had been sighted all around us. Earlier, I began to bring our valuables downstairs in case we got a bad storm. I was getting really panicky!!

By now, it was about 8:00 and I was nearly out of my mind! I heard on the radio that there had been a funnel seen over the area where my mother, sister and her family lived in St. Paul. I couldn't stand it any longer so I brought Pat, Chris, and Philip downstairs with me while Duane and Mike stayed upstairs watching the storm. I brought down a Blessed Candle that my mother gave to us for luck and lit it. Just then, I heard Duane and Mike come running down the basement stairs and I knew we were going to get it. He shouted, "I can hear it coming, everyone under the stairs!" The kids started to scream as I grabbed them. We just got under the steps when it hit! The time was 8:52! I can't describe

the sounds! The lumber crunching, the chimney bricks falling, windows breaking and that horrible wind! I had my arms around the kids and I could feel the suction on my arm and dress sleeve as if the wind was trying to take me with it. Then, there was dead silence. The steps had fallen on top of us and Mike was holding it up with his back. Duane stepped out and looked around and started to laugh. I was afraid he had got hysterical and then he said, "Georgia, it's gone, everything we had is gone!"

We got back to ground level and we heard voices calling out, "Oh, my God, where is my baby?", "John, where are you?", "Someone help me!". Then, our neighbor lady called for help. We ran over and she was in her basement and her car had fallen on top of her! She put up her hands to us and literally walked up the wall. Her face was a mass of blood! It looked like someone had poured a catsup bottle over her head. Our house was gone and we didn't have any bandages so I tore off one of my dress pockets to put on her head, which got soaked in no time. We had to get her to a hospital and no one would stop. Finally, Duane stepped into the road and stopped a kind gentleman who would take her. Our three smaller children were shivering from the cold wind, frightened and crying as you've never heard a child cry!

Just then, a man came running down the street and said that another neighbor was badly injured and someone had to sit with her until help came. I was the only one available so I made the kids link their arms together and told them to stay in our front yard and not go anywhere until I got back. The lady was lying in a pile of bricks, broken lumber, and insulation. She was a widow with three children and she wanted to know where they were. She had gone to bed earlier and was blown right out of her house into the front yard. It turned out that her kids were thrown out of the second story bedrooms into the front yard as well. They had minor cuts and injuries. I could hear the sirens start to howl and it seemed like they were coming from all directions! So there I was, sirens howling in all directions, not knowing where my husband and children were at the time, without a flashlight, sitting in the cold rain with possibly a dying woman. I wanted to pray with her but I was afraid I'd frighten her, so I kept my prayers silent. I just talked, I talked about the flowers I would plant that Spring, I talked about the pretty clothes she had, I just talked to keep her mind off her pain.

A Civil Defense worker came up to us to check on her condition and the police came with a stretcher and took her to a waiting ambulance. I found my children sitting on a couch at a neighbor's house, covered up with blankets, just staring off into space. They too, were in a state of shock. People kept running in and out of this house. Although there were no doors or windows, just a roof, to us it was a haven! People kept asking if their children had been there or had been seen by anyone. It was heartbreaking to see the look of anxiety on their faces. Finally, a friend of ours, who lived nearby but wasn't hit by the tornado, came to us to ask what he could do for us. Those were the sweetest words we could hear at that time! We asked our friend to please take us to my mother's house in St. Paul. As he drove us down there, it poured all the way and not a word was uttered by anyone unless it was for directions.

The next day began the heart-breaking job of going back to pick through the debris to try to find something that wasn't smashed or soaked with rain, something to help start over again. It was futile. Everything was totally destroyed. We learned then that 74 homes were destroyed, condemned, or damaged in our area that night with 5 people dead on our block alone! Two days later we were told by the schools that our children had to finish their school year even though there were only 2 weeks left in the school year. We knew it would be almost impossible to drive them back and forth every day from St. Paul to Mounds View. Duane's sister, his nephew and wife generously offered to care for our children and send them to school in their town which was 75 miles from St. Paul. There were 6 pretty upset people when we drove off back to the city leaving the four of them behind.

We decided to rebuild in that neighborhood. We were the first to move there in 1953 and we were the first to come back after the tornado. By July 29, 1965, we were back in our new home! We had bought new furniture, clothes, and a used car. We were a family again because we were under our own roof again! It couldn't have been accomplished in such a short period of time without the help of my sweet mother, Mrs. Helen Thole. But, I have to admit that in the spring, when the skies get dark and we hear the possibility of tornadoes in Minnesota, our hearts beat faster, our hands get clammy, and there is anxiety in Mounds View until we hear the "all clear" siren. One day when discussing

the storm with a neighbor, Duane asked, "Why did God do this to us?" The very wise neighbor replied, "God didn't do it, it was the work of the devil." We heartily agree!

Chris (Buck) Hodges, 2303 Lois Drive, Mounds View

(11 years old at the time of the tornadoes)

I remember that it was really stormy outside, more than any other storm I can recall. There was a tremendous amount of lightning and we were downstairs playing cards in the basement. My dad and brother, Mike, were outside watching the storm. All of a sudden, they came running down the stairs and yelled at us to get into the well pit under the stairwell. We all tucked under the stairs when the lights went out and I heard the loudest roar that I'd ever heard. As the house was being taken, it broke the water pipes over our heads and dumped ice cold water on my sister and me. We were screaming but you couldn't hear it over the noise of the tornado. The time was just before nine o'clock in the evening. Then, it was over. It was dark and dead quiet.

You could see the lightning on the other side of the basement due to the floor that was lifted up from the foundation. My dad went over to look and became a little hysterical saying that "it's all gone, it's all gone..". The steps had fallen down and my dad and Mike went over to jiggle the stairwell back into place so we could all get out of there. We ran up the stairs and the house was gone! We just stared and were in shock. There was nothing left. As we looked around, all the other houses around us were gone as well. That's when we heard our neighbor lady calling for help. She was trapped in her basement with a car that had been flipped over in it. When we got her out of the basement, her face was covered with blood. Dad made Mike take off his shirt so that the lady could stop the bleeding and clean up a bit. People were just milling around in shock. No one said a word. Everyone was staring off into space and it was very frightening. Our parents went off to help other people and told us to stay put and don't go anywhere.

It was cold and rainy. Our neighbors in back of us, the Whitenecks, came over and had us go over to their house. It was one of the only houses in the area that you could get out of the rain. This became the neighborhood gathering spot, people would come in looking for loved ones that were missing and see if they had

been there or not. I remember one man named Abraham coming in looking for his family and we later found out that he lost his wife and two daughters in that tornado. Soon, our parents came back and we got a ride from a gentleman down to our grandma's house in St. Paul. We were shipped out to a small town to finish the school year which had 2 weeks left.

We were the first ones to move back into the neighborhood when the house was rebuilt. There were piles of rubble everywhere. No houses, flat basements. I know that I spent many a night sleeping in the basement after those tornadoes. A couple of strange things happened in the storm that night. My dad found his welder that had been in the detached garage, in a neighbor's tree down the block. Our camera/projector was jammed full of cheese sandwiches that we had made for school lunches earlier that night. I never saw such lightning since that night! It was as bright as day when the lightning would flash and it seemed ferocious in nature. We never did hear the sirens that night due to finding a piece of wood that had been lodged in it during the storm. I will never forget that night.

Paul Popelka, Route 3, Glencoe

I remember that the tornado started out south of Glencoe and moved north and hit the east side of Glencoe and moved further from that point. Our farm was about 1/2 mile east of the damage path. The tornado didn't really look like a typical one. The skies were white, blue, with stuff flying around in the clouds. The farms out here were hit pretty bad and the tornado wiped out the John Deere dealership on the edge of town. We had "tough" weather that entire year. It didn't surprise me that we got the bad weather. The previous winter gave us a ton of snow. The spring had a fast warm-up which led to massive flooding. The mayor of Glencoe authorized dynamiting of a bridge out here to help relieve the flooding in the town of Glencoe but that went poorly. The storm happened really fast but the thing I remember the most was that everyone went around seeing if you could help each other. You just salvaged what you could and made the most of it.

Tom Rockvam,
1-3rd Street, Excelsior

(personal recollections including excerpts from his writings)

Every spring since May 6, 1965, it has bothered me as to how and why the tornado that demolished the western end of Lake Minnetonka took the path that it did. I owned and operated Crepeau Dock/Sales & Service and the ice out that year was one of the latest in recorded history. We were putting in docks that day and it felt very strange all day long. It was really hot and muggy and as the day wore on, the sky became more green. Usually the phone rings off the wall during the spring but this year, the only time in over 30 years, it didn't. After the tornadoes came through, everyone was helping one another to get their lives back together and there wasn't time to launch boats, etc. Before I begin with the story, I want to mention a strange tale that relates to the 1965 tornadoes. In around 1998, St. Peter was hit by a terrible tornado and ever since 1965, whenever Minnetonka won a state championship, a big tornado would come through. It got to the point where people around here didn't want Minnetonka to win otherwise we or someone else near by would get nailed by a tornado! So this tornado hits St. Peter and there is a gentleman working in his yard in Wayzata and he finds a cancelled check from St. Peter dated May 6th, 1965! If that isn't eerie, I don't know what is! That story about the cancelled check along with the actual picture was in the Lakeshore Weekly but I'm amazed that they didn't mention the date on the check! Now, back to the story.

The tornado started out west by Norwood-Young America and traveled in a normal north/northeast route as it touched down, devastating Loring Acres, a new housing development on County Road 44 in Mound. It then continued going north/northeast, hitting the end of Hardscrabble Point, again causing considerable damage. The tornado then continuing on the same course headed across Cook's Bay of Lake Minnetonka towards the north shore of Cook's Bay, Shirley Hills School, and Tonka Toy's main plant in Mound. It was here that the tornado went awry! Why? As it was coming across Cook's Bay, it kept sucking up water.

According to people I talked to who lived on Cook's Bay, they had sand beaches extending out up to 200 feet further than normal. Then just as quickly as it went out, the water came gushing back past its normal level and traveled 100

feet up the shore, depositing fish, trees, building materials and other debris on and in their homes and property. It resembled a mini-tidal wave caused by the tornado.

At this same time, the tornado took a drastic turn to the right, 70-80 degrees, and smashed into the west shore of Island Park, killing people and destroying everything in its path. I wondered if the extreme flexing of the lake level contributed to the tornado turning so drastically to the east. I sent an email to head meteorologist Dave Dahl at KSTP Channel 5.

He wrote: "That tornado event was probably the reason I got into meteorology. I was 10 years old at the time and was fascinated by the three separate lines of storms that caused tornadoes that night. I've studied tornadoes for a long time and have realized that the ground (or water) they travel over has a lot to do with their character. Even though the parent thunderstorm eventually determines where the storm travels, short distances can be influenced by the terrain. When a vortex travels over water, the inflow is almost uninterrupted. The rotation easily maintains itself, while the speed decreases due to added weight within the rotation as water is drawn upward into the vortex. Compared to other debris, such as trees, building materials, and just plain dirt, water is much more dense. This has a tendency to slow its forward progress, and abruptly change its direction of motion. It could have easily caused the extreme right hand turn, because as the parent thunderstorm continued to move, the vortex slowing at all would mean an eventual right-hand turn to "catch up" as the meso (middle) of the tornado continues to rotate. I hope this helps explain why the tornado took such a drastic right turn."

Like I said earlier, the conversation I had with the people who lived on the north shore of Cook's Bay told me about the immense fluctuation of the water at their lakeshore. Then there was the woman I talked to who told me that she had pulled off County Road 110 by the Surfside (now Chapman Place Condominiums) during the storm because she couldn't see to drive. All of a sudden her car and all of the land around her was covered with fish, tree limbs, building material and gushing water.

After slamming into West Island Park, the tornado receded back up for a little bit and coming down hard hitting the east side of Island Park from Black Bay to Spring Park Bay. It went across Spring Park Bay and hit again, this time into Casco Point, it leveled many homes on Casco Point and then lifted for a short distance and came back down and completely wiped out the south side of the town of Navarre along with the houses on the north shore of Carmen's Bay. I would say it took 10-15 years to rebuild Navarre to any semblance of what the town used to be with all the little quaint shops in the area. After it left Navarre, the tornado went back up for a short time and then came down in the area just before the Arcola Bridge in Minnetonka Beach and totally demolished homes in that area before lifting up, joining other tornadoes on their way up to the Fridley area.

At the same time, there was still another tornado that hit Chanhassen and did a lot of damage in the Christmas Lake area, lifted back up and passed by the east of Excelsior and came down in full force and wiped out the Minnetonka Manor and Cottagewood areas. After it left Cottagewood, it sailed back up, crossing over Wayzata and proceeded to hit the northern suburbs.

Betty & Robert Johnson,
South Arbor Lane, Minnetonka

We lived just off Shady Oak Road between Highway 7 and Minnetonka Blvd. We were in the walkout basement and went outside and looked straight up to see that tight swirling mass going past overhead. It didn't touch down in that vicinity. You could see the rotation just above the trees! It was really foolish of us even being out there but my husband Bob was a scientist and we heard about the storm over the radio. It took about 2 or 3 days later before we went around to look at the damaged areas. You heard a lot about those storms the next few days. In 1968, we moved to our present house on March Circle, near the intersection of Plymouth Road and Cedar Lake Road, north of Minnetonka Blvd. Two huge oak trees very close to our new home had been topped when one of those 1965 tornadoes dipped down a little lower there. You could definitely tell it was tornadic even after three years.

Robert Christenson, 231 Rice Creek Terrace, Fridley

Here is the historical events as to what my family and I endured the night of May 6, 1965.

6:00 p.m: I was at the Fridley Commons ballpark playing ball and it looked like rain and overcast.

6:40 p.m: It began to rain and I left to go home. On the radio, I heard warnings about hook echoes in the Chanhassen area and that the tornado was heading in our direction. So, I went back to Fridley Commons to warn the people that were still there.

6:45 p.m: I arrived at home. Mom, Dad, Russell & Leona Olson were playing canasta. After telling them what I heard on the radio, Leona said, "We'd better leave!" It started raining hard.

6:50 p.m: Mom said, "How can we have a tornado when it's raining?"

6:52 p.m: Rain stopped. Very quiet. On the radio, WCCO states that the tornado is heading for the NW suburbs. The western sky was pink colored.

6:54 p.m: Hail starts and becomes baseball-sized! My brother John takes my dad's car

out of the garage and puts his in!

7:00 p.m: Hail stops. I had my binoculars and looked out the window because WCCO stated that the tornado was at Hwy. 100 & Golden Valley Road. The western sky turned from pink to black. White, spider-like clouds were flying to the south at a fast rate.

7:03 p.m: Sirens sounding in Minneapolis. The fast moving, white clouds going over us were maybe 100 feet off the ground. It looked like small puffs. I looked to the SW and saw a cloud with a tail.

7:06 p.m: Became very dark overhead. A big cloud came out of the sky with a roar! I yelled, "Everyone to the basement!" But, my dad wanted to stay and watch from the kitchen window. My brother and I pulled my dad downstairs and I ran to my bedroom located in the NW corner of the basement. There was a vacuum feel to the air!

7:07 p.m: Power out! After about 30 seconds, the roar was gone.

7:08 p.m: From dad's office in the SW corner of the basement, we could see that the windows in the 3-season porch were blown out.

7:09 p.m: We went upstairs and where my

dad had been standing in the kitchen, the window had been blown out! I looked at the debris, our gas yard light was still standing in the front yard. I looked to the south and saw another tornado coming, a solid column with debris being wrapped up in it! This tornado was much louder! We headed for the basement again with our ears popping and we stayed in dad's office this time. It lasted longer, about a minute.

7:12 p.m: Noise subsided and we went upstairs. The other kitchen window was gone and I looked out front and the gaslight was gone! The wind was still blowing hard!

7:14 p.m: Debris everywhere! The strange event that this tornado did was in our front yard. The tornado stripped the telephone wire off the pole and wrapped it into a volleyball-sized ball of multi-colored plastic hanging from the telephone wire! Most of our shingles were gone and we talked to several neighbors.

7:18 p.m: Went to my Grandma's house off of Mississippi. John took my dad's car which had a broken window. I took my 1964 Thunderbird. When we got there, my uncle, Waldo Oman was laying on the ground. He had tried to open my Grandma's side door when the tornado hit. It created such a vacuum that my uncle couldn't open the door and the tornado picked him up and spun him around the house to the opposite side. All this time, my Grandma never knew what had happened to Waldo.

7:30 p.m: Sgt. Fritz of the Fridley police deputized me and placed me at the intersection of Mississippi & University. I was to stop all traffic going north on University and east on Mississippi. He gave me a bunch of flares to use being that it was quite dark out now and that all the power was off in the area. I sent everyone to City Hall that was injured, police from other municipalities, and flares from anyone who had them. It didn't take long for the looters to come out. In fact, we apprehended a "makeshift" Red Cross truck that some men were using to pickup TV's, etc. They had painted a big Red Cross on the side of a truck and almost got away with it! One day before the tornado hit, the Red Cross center at Redeemer Lutheran Church on Mississippi closed after the big floods of April.

8:00 p.m: WCCO states that another tornado is heading toward us! I simply could not believe it! My best friend, Philip Polivchak, was coming to see how I was doing when the tornado picked up his car at Hwy. 694 & University. After picking up his car and dropping it, he spun

a U-turn and got out of there! I looked SW and saw the huge tornado. There was green, yellow lightning flashing inside the gigantic column! It smelled like sulphur out and yelled at everyone to go west on Mississippi to get away from that tornado! I dove under the bus bench and hung on for dear life. It was made of concrete and heavy duty. The tornado came within 4 blocks of us but didn't hit us directly. It was like being sandblasted. The noise was so loud, you couldn't hear yourself think! After the tornado had gone by, I went to City Hall and got a yellow raincoat due to so much rain. My duty ended at 5:00 a.m. the next morning. I shudder when I think of all the smoke I inhaled during that shift from the flares.

After I left my duty, I walked over to the Rice Creek Terrace area. The devastation was incredible! I remembered seeing a house that was gone, just the floor left. But, there still stood the table/ chairs/ plates/ condiments/napkins! The tornado had not even blown away the napkins! But, at the same time, totally destroying the house! It was an evening I will never forget!

Willis Erickson,
7341 Memory Lane, Fridley

I was at a dinner-meeting in downtown Minneapolis when the tornado tore through the Fridley area. A business associate had a call from his wife in Brooklyn Center who informed him of a tornado there and that there were reports of others in the area. An announcement was made in the meeting and those of us who lived in the area left the meeting to try and call home. The phone lines in our area had been installed underground and I was able to get in touch with my wife. She was crying when she answered. When asked how everyone was, she replied that she and the two small children at home were fine. She had grabbed a mattress, bundled herself and the kids into a crawlspace under the foyer of our split-level home, and pulled the mattress over them. She told me that just after the noise subsided, the phone rang (it was me) and she went upstairs to answer it in the dark but did not know at that time whether the house was intact or not. She was very concerned about our third son who was in Spring Lake Park taking an accordion lesson at the time. I told my wife to hang tough, that I would try to get home as soon as possible and would first try to get to Spring Lake Park to check on our son there.

Upon leaving the parking lot at Charlie's Café in downtown Minneapolis, I was accosted by the parking lot attendant who informed me that traffic was banned on the downtown streets and I was to stay where I was. The attendant stepped aside when he was informed in no uncertain terms that if he stood in my way he might get run down. There indeed was no traffic in downtown Minneapolis until I reached Central Avenue and started north. Then the "fun" began. Proceeding north toward SLP the driving became more hazardous with each passing block. When I got as far as Columbia Heights there were utility poles, electric lines, and much debris scattered over the avenue. I don't recall how many times I led or followed a line of traffic thru the median strip ditch in order to get to the other side and back again. At one point, there was a city bus stuck in the median strip and I helped many others to push that bus out so traffic could continue (ruined my good business suit in the process). It was an eerie feeling while making this entire drive because of the lack of electric lights anywhere.

The music studio in SLP had been demolished and there were just two portions of

walls standing at a stairway to the basement. While pointing my flashlight around the mess where the building had been, a lady appeared at the top of the stairway. I will never forget her words, "Oh, Mr. Erickson, your son Steve and all the kids are safe in the basement!" This courageous music instructor had her charges into the basement where they were drinking pop from a machine down there and some were even playing their accordions!

After loading Steve and his accordion in the car, we began the adventure of trying to get back to the Melody Manor area. We first went south on Central Avenue but could not go west on 49th Avenue so backtracked north to Hwy. 10 and west to East River Road. Traveling south on East River Road it was very difficult to negotiate because of downed utility lines and trees. All streets going east were impossible to travel on and we were forced to go all the way to I-694 where we drove back east to go north on University Avenue. We got as far as Holly Center, University & Mississippi, parked and locked the car, and began walking north on University. Avoiding downed electric lines and other debris, the walk to Melody Manor became more treacherous as we proceeded. I don't recall seeing anyone the mile or so between Holly Center and our home. The closer we got to our neighborhood, the worse things began to look. Can't recall the exact time but it must have now been nearly 3 hours since I had left Charlie's Café.

It had been raining for some time. When we reached home, though it was very dark, I could see that homes just across the street were no longer there. The only light in the area was from a dome light burning in an upside-down car in a front yard across the street. After determining everyone had no physical injuries, I began to look for things to make us more comfortable. The garage was intact and I got our a Coleman lantern, lighted it, and hung it in the family room downstairs. All upstairs windows had been blown inward and I found enough storm windows to cover those openings. However, it was still raining and coming in through large holes in the roof. Two doors south of us had been a model home. I went there and found a roll of tar paper in the basement hole, crawled up on our roof with it, and covered the holes to keep water from pouring into the attic.

By this time, many neighbors, attracted by the light from the lantern began to wander down

to our family room. Many looked like walking zombies, what with trying to recover from the shock of missing and damaged homes. We broke our refreshments for those who wanted them. During the time we were gathered there, several rescue squads appeared and were checking to see if all in the neighborhood were accounted for. Before I got home, a neighbor who lived directly across the street and somehow managed to get home right after the tornado hit, had helped his critically injured wife to an ambulance which had been able to get within a few blocks. In the meantime, nobody could account for their small son who was missing. While we were out searching for him, an announcement was heard on WCCO radio that a lady who lived several blocks away had seen the youngster and taken him in. When the tornado struck, he was downstairs with his mother and said afterward, "My mommy won't talk to me". Apparently, he had crawled out of the rubble of what had been their home and was wandering around when the lady spotted him. Another neighbor lady had been sitting with us for some time when one of the rescue squad members determined she had a broken arm.

She was taken to the hospital. Miraculously with all the devastation (39 homes leveled) in this neighborhood, there were no fatalities!

Sometime during the night, the National Guard was called to police the area and I heard of no looting. I worked at the Onan Corporation and first thing the morning of the 7th, went to the plant and picked up a generator. At least we had partial electricity and had a coffee pot going for any who wanted. There were 3 homes left standing on our side of the street but nothing for a quarter-block in either direction. Facing us were 39 leveled homes. Madsen Park is at our back yard and all manner of debris was scattered throughout the park. I got up on our roof and took photos of our section of Melody Manor that had now become a "no-mans-land". Other areas of the metropolitan area were probably hit as hard as we were but the Lord blessed us in that there was not one fatality in this part of Melody Manor!

Robert J. & Barbara Hughes, 548 Rice Creek Terrace, Fridley

We moved to our home in Fridley in 1959 and on May 6, 1965, we heard the warnings for tornadoes heading in our direction. We looked at the green skies and I (Barbara) knew what tornado skies looked like growing up in Indiana. We all went to the basement with our three children (ages 8, 6, and 2) and shoved the kids under a big, old desk we had in the basement. I huddled over our two-year-old. We had just finished dinner and the tornadoes didn't give us time to even clean off the table! From the basement window, we saw our "burning barrel" rolling in our backyard with sparks flying all over the place. I was afraid that the barrel would hit the house and start a fire! The roar of the tornado was terrible, sounding like the classic freight train sound that everyone compares it to. We were very fortunate losing only some siding and the windows on the west side of the house. There wasn't much time in between storms and I remember going with our family to the basement again when the other tornado came through.

So many families were divided that night due to so many activities and this caused a lot of stress not knowing if your family was OK or

not. We happened to all be together that night because Bob stayed home instead of going to a scheduled meeting but due to the weather, decided to stay home with us instead. Our kids remember that Bob really stressed wearing shoes due to all the debris in and around our house and he had to dig around upstairs to find 3 pairs of shoes for the kids and that flashlights were needed due to the lack of electricity. When the tornadoes hit, our toilets were flushing due to the sucking power of the storms!

After the tornadoes, our neighbor Morris Myer (a member of the National Guard) came over with a much needed generator for us and about 5 other neighbor families to keep one appliance going in each household. At least we could keep the refrigerator going. You could only plug one item in at a time, for example, if you wanted to vacuum clean the house, you had to unplug the frig in order to do this. I taught Sunday School and for at least the next year or two, whenever you talked about storms, the kids would stop and really listen. Typically, those kids are real squirrelly but not when you mentioned "tornadoes". So many people were underinsured in those storms and that caused a lot of misery. A strange outcome of the tornadoes was

Lorraine & Floyd Ordemann's house on 67th Avenue NE. Their roof was picked up and the curtains were flipped over the exposed wall and the roof set back on top with those curtains flapping in the wind!

John Hansen,
5809 Tennison Drive, Fridley

(John was Superintendent of Fridley Schools at the time of the tornadoes)

We lived on the east side of Hwy. 65 on a hill facing west and I had just dropped off two of our children at the Fridley Junior High for a science fair going on that night. The news of TV told of the coming tornadoes and at 7:06 pm, I a big cloud of black was coming into the area of Fridley Senior and Junior High Schools. Roofs and other building materials were flying in the air! It was very fortuitous that all the activity buses had brought the students home and that those tornadoes came at a time of minimum exposure to our students, parents, and teachers. Most of the day's activities were completed or to happen later. The tornadoes damaged all the gymnasiums at the Senior and Junior High Schools. Total damage of the schools was approximately 1-2 million dollars. At that time, we could construct a school for about a half million dollars.

Some important background information about that situation should be mentioned in terms of insurance and how quickly we were able to begin running classes for our students. When I arrived in 1959 as Superintendent for Fridley Schools, we were a rural elementary school district. We had an insurance study completed to assess future demands and found that a replacement-cost plus basis plan to be the best for our district. This provided protection of full inventory, supplies, and replacement of buildings. This was of tremendous benefit to our district when the Fridley Tornadoes destroyed/damaged most of our schools, resulting in the quickest, easiest insurance claims in the history of Minnesota schools. The district wound up ahead due to this good planning. For example, Hayes Elementary School was constructed by the same contractors that had built the earlier version only to be built bigger and better in less than 6 months. This was also one of the fastest school constructions in Minnesota. Occupancy of Hayes was in December, 1965.

After the storms hit, there was a tremendous cleanup of the schools by our custodians, teachers, parents, and students. I was impressed by how so many people came together to help get our schools running again. Everyone was truly at their best in this situation! We were able to have students return the next Monday for classes and we finished out the school year. To my knowledge, there was only one severe injury and that was at Parkview Elementary when a cinder-block fell on a woman. When the tornadoes hit, our teachers were already familiar with Civil Defense and knew where to put students and parents. They got everyone out of the gymnasiums and into the better constructed hallways. This is the reason why there were so few of injuries and fatalities in the Fridley schools. Students were kept until the parents arrived to pick them up. After that, police and fire helped bring students home to which many didn't have anymore in Fridley.

I arrived shortly after the first tornado came through at 7:06 pm but felt that there were two other tornadoes that came through that night. Each time we ducked for shelter and luckily no injuries. My new 1965 Plymouth had the paint blown off by the tornadoes and none of the windows were broken. Inside the car, there was tar and dirt all over that had been blown around the window seals. Our home only suffered some shingle loss and one blown in window which really frightened my daughter. The two biggest

items I remember about the tornadoes was how well everyone came together and worked so hard to get our schools going and how well we were covered by insurance and very organized to get our schools repaired and rebuilt.

Eleanor Rittenour,
Fridley Terrace Trailer Court, Fridley

Suddenly it was quiet, cool and moist, and the smell-sweetly acrid, thick and unforgettable. I was staring at the gray, guilty, yellowish sky. It was May 6, 1965, and our lives had changed that moment with nature's deadly, but natural crime.

Our daughter, a toddler, would be two years old in nine days and she was sleeping in the back bedroom. We had finished supper, the dishes were washed and I was preparing to visit a friend for a haircut, but was delayed because of the thunderstorm taking place outside. My husband, Dale, was home. It was hot and I was getting impatient.

The wind began to blow slightly, it started to rain and we saw our neighbors to the east of us two lots down were dashing out to their station wagon-mom, dad, and three children-none of them had their shoes on!

When all of them were in the 'wagon, the driver made a hasty U-turn and off they went. We noticed the occupants in the windows of the mobile home on the corner across the street. By now, not only heavy rains were falling, it started to hail. Small and pea-sized at first. We watched a man from that mobile home dash through the hail and spread himself with his jacket extended out across his car hood to protect it from hail damage. In ignorance, we chuckled at the sight. Eventually, the hail grew to golfball-size and became too fierce; he retreated back to his mobile home.

It started to let up (at last I can get to that haircut appointment), it grew quiet, the wind and rain completely stopped, I opened the door to our trailer and peeked out and then stepped onto the small, open, wooden deck to survey what we could see of our neighbors and the park. As I looked to the west, I saw a number of bright, blue flashes coming from a large industrial plant which I speculated was Northern Pump Company.

My eyes traveled from those flash points upward and easterly until I was looking just a few degrees less than perpendicular from our trailer. For the first time in my life, I saw a mushroom-

ing, rotating circle of clouds, boiling and writhing, almost directly above. Still I heard nothing. I yelled to Dale, "it's a tornado!" and ran to the back to grab Becky and laid her on the bed in the middle bedroom and I put myself on top of her!

In the meantime, I heard Dale say he was going to stay at the doorway to watch what was going to happen-if anything. In Less than ten seconds, he flew onto the bed on top of me and held onto the mattress to secure us all down.

The roar began, like a train bearing down close. Things began to shake; suddenly we were lifted into the air (about 5-6 feet) and swiftly slammed down, still upright. Every window in the trailer was broken as I listened to glass shattering everywhere.

As we lay in the bed hanging onto each other with Dale clutching the mattress, we seemed to be in the "eye" of the tornado. This unspeakable natural power produced the most helpless feeling I had ever experienced. Nature's madness became quiet, calm, silent. I had enough time to ask Dale, "Our home-what are we going to do?" His reply was, "All we can do is pray." To myself, I was already saying the Lord's Prayer-no time to create a prayer, my thoughts could only bring up the rote.

Then, it seemed the back half of this great storm hit. Again, it picked us up. We rolled and tumbled. Everything was out of control and then it was done.

We found ourselves smelling the tornado's devastation, lying on our backs on the ceiling of the trailer. Still piled over each other, still hanging onto the mattress. For a moment, we were in shock. A feeling of cool, blue mist crawled over my face as I looked up into the sky.

I looked around and noticed the floor of our trailer was turned over and straddling the back bedroom of our neighbor's trailer. As we climbed out of the debris, there was a large, at least eight-foot diameter steel cylinder, looking like a septic tank, rolled into the front of our lot. Still under the spell of that horrific act of weather, stirred up by hot and cold winds, we called out to see if anyone was still left, or even alive, in their homes. It seemed we were the only ones who had not abandoned the park to seek shelter. No one seemed to be around. We took notice of ourselves and checked the bleeding from a superficial wound that appeared on Dale's lower leg. It wasn't until later, after noticing a sharp digging and itchy feeling on my chest, about a tablespoon of glass from some smashed light

bulb had accumulated in my clothing and was piercing my skin as I pressed my daughter's body against mine.

Things we owned were scattered on the grass. Luckily, much has been held together by the now swaying walls of the upside down trailer, still somewhat aligned north and south. A utility pole about five feet high, located on the east side of the trailer was now on its west side. We decided we must have been lifted and relocated at least that height to have been flung over the pole without damaging it. It was impossible to know if we tumbled over more than once. The floor, with its wheels attached was straddled, upside down, the back of our neighbor's trailer tottered east and west. If it hadn't been for the piano holding parts of the floor to the side we could have been crushed, the refrigerator had soared over us and landed in the bathroom!

We were still trudging down the curved street, holding our two-year-old, still calling out, finally getting to the office and Laundromat facility and found where all the residents had gone. It was wall-to-wall people. By now it was getting dark. No electricity-no lights. Astoundingly, there were Civil Defense workers standing on the street around the curve! It seemed incredible they were responding to this disaster only minutes after the event. We nestled ourselves against the north wall in the park Laundromat. We heard the sounds of crying, fear, and desperation. It was hot and stuffy. I could smell the sickening odor of vomit of some child who must have eaten a peanut butter sandwich earlier. Again, we were jerked into reality as someone yelled, "Here comes another one! [Tornado]". The doors had been open. Women called out to close the door. Another called to leave them open. We were all lying on the floor, barely room as we all tried to flatten ourselves to the floor for protection! This time that tornado passed over us without touching the ground. Relief descended replacing the panic and fear which dominated that night. Future storms would cause those feelings to descend again.

A woman was brought in on a stretcher and she was badly injured. She had blood all over herself. They laid her on one of the tables. I had grabbed a blanket, a wedding present, as we left the trailer and put it on the trembling, moaning figure. Another voice out of the darkness called out, "There's some strangers already trying to steal our stuff" in the trailer court. Many of the men, including Dale, gathered and went out to patrol hoping to scare the looters away. I had found a chair to sit on and was still clutching our daughter.

The years following our experience both Dale and I experienced a primitive sense of fear with every thunderstorm that presented itself. We would find ourselves pacing outside watching the skies and listening to the weather reports. It was at least five years before that awful fear began to leave. We were awarded $2300 for our trailer from the insurance company which became the foundation for buying our first home two years later. After the tornado, we settled in a basement apartment, which was difficult to find because of the many displaced families also looking for shelter, I became sick with various maladies for several weeks. I became aware I had experienced whiplash on the left side of my neck and that pain has been my companion since.

The tornadoes of May 6, 1965 changed our lives forever. From that experience, while it took time to develop, I was able to understand the insignificance of material goods in my life-that we had the strength and endurance in us to build again a place to call "home". As long as we held ourselves together as we did on that mattress, despite nature reducing us to nothingness in its power, becoming little more than the pieces of debris flying in the wind, we survived the very brush with death. Something that fearful actually morphed into an appreciation of the true, priceless priorities in life and the ability to be able to sort what is genuinely valuable!

Don Watson, Spring Lake Park High School, Spring Lake Park

(Don was the Athletic Director of Spring Lake Park High School at the time of the tornadoes)

Students and teachers had been sent home because of storm warnings but our baseball team and track teams were yet to return home after having competed elsewhere that afternoon. I remained at the school to warn the coaches and athletes of the possible storm and waited with two students who lived in my general neighborhood (south and southeast of the high school) until all but we three and two custodians remained.

As we were just about to leave via the exit where the Fine Arts Center is now located (the south exit near all of those glass windows at the south side of the large gym), it began to hail. So we tarried. It was a brief hailstorm but

some of the stones were baseball size. One of the students collected one which we were going to put in the freezer and show others the next day. One of the custodians volunteered to put the hailstone in the freezer on his way out but it never made it into cold storage!

Suddenly it became awfully quiet. We looked toward Fridley. The storm struck! The tornado was so wide and green, we had a different perspective because it was so close and on the ground. It didn't look like a typical cone-shaped funnel. We had time to see the wood constructed football tower disappear. This tower, constructed by staff, had four 4x4 legs embedded in the ground with cement. We later found three legs with the cement intact. The fourth leg we never found.

At this point we all headed for shelter. Two headed down the hallway. One custodian sought safety in the boiler room. The other narrowly missed being crushed by a falling roof in the old boy's locker room.

The two boys and I sought refuge in the brick corner opposite the southwest gym door. From our "safe" corner, we viewed what a tornado does to a strong building.

We saw the windows puff in and out and disintegrate into millions of pieces! We saw garbage cans "float" up and down the hallways but not at the excessive speed. What a way to get rid of garbage! We felt something strike our wall haven which shuddered but remained intact. It became lighter inside as a portion of the roof disappeared. We heard the "freight train" sound of the tornado but through it all we weren't frightened or injured!

After the storm passed, we became frightened when we saw the devastation to our school and to the neighborhood to the south of the school. Our first concern was for the families who lived in that direction. I warned the boys to be wary of fallen wires and we separated to search for our families. (They were all safe!)

The next day when I returned to the school in the daylight, the area looked as if it had been in a war. Damage was extensive. Walls were down, roofs were gone or leaked and windows were broken. Yet in it all, there was some humor. I can remember a perfectly good fishing boat leaning up against the wall on the inside of the typing department. The other wall of that wing which faced south out toward the football field was completely demolished. It appeared as

if someone had carried the boat into the classroom and learned it up against the inner wall. The gymnasium floor was soaked from the rains that night. The wood swelled which had caused the floor to heave upward into a huge mound. The athletic storage area was intact but the roof leaked. All of our gear was wet. Students and I hauled this gear to our new house which was under construction in Andover. We hung the uniforms and gear over the bare rafters to dry.

Maude Ruth Watson,
7811 Jackson Street NE
Spring Lake Park

(Wife of Don Watson)

Thursday, May 6th, had been a very busy day. David was to sing with the sixth grade chorus at Woodcrest PTA at 8:00 p.m. and was supposed to be at school by 7:30 p.m. so we all were rushing to make that. I dashed home from work at 5:00, started the washing, started supper, the boys showered and dressed in their good clothes, one line full of clothes got dry and the last ones were hung out when it really started getting cloudy. It had been hot and windy for two days, humid, Don said it would rain because his knees said so! One girl at the office who has arthritis quite badly hadn't slept for two nights, and I had such a sinus headache, that we all remarked we hoped there would be a change in the weather.

Our daughter, Donna Ruth, hadn't showered yet and was still excited because of cheerleading tryouts that day (she made it). She kept remarking that she was hungry but we wanted to wait until Daddy got home from school. About 6:55 p.m., it started to sprinkle so Dan and Christine (neighbor girl) helped me bring in the clothes off the line. The kids were watching TV, and as I checked on supper and dashed through the kitchen to check the clothes, there was a news flash on TV about funnels sighted in the Lake Minnetonka area, southwest of Minneapolis.

I turned on the radio to WCCO right away and we had them both blaring. Donald called just about 7 o'clock and said he was ready to come home, but by then it was raining good and hail about 2-3 inches in diameter was falling. One crashed through the storm window on the northwest bedroom (because it was anchored out, the old-fashioned kind) so I dashed in to close that. I told Donald I would pick him up just as soon as the hail let up and to stay there until I got there.

The radio said a funnel had been sighted traveling toward Osseo so I figured that the one from Lake Minnetonka was taking a pretty straight northerly course. About that time the kids mentioned again that we should eat, supper was getting cold, so I told them to fill their plates and take them to the basement while I went to get Dad. We turned off the TV and unplugged it, the kids went with their plates downstairs, and I picked up my purse and started out the breezeway door. All this time I had been walking from window to window trying to sight what I thought was a funnel, but never did see one. The clouds were black and green (which Mom always said was hail) and were traveling pretty fast toward us. Probably what saved me was that the wind started to blow as I reached that door and suddenly I decided to wait. I came back up, unplugged the radio, turned off all electrical appliances, lights, took my purse, and tucked the radio under my arm as I went downstairs. The back door banged open, glass broke, and things started traveling past the basement window. Large objects whizzed past the window and in the darkness we couldn't tell what they were.

We thought the top of the house went because of all the banging. I thought the car rolled past the window but it didn't. The fuel oil barrel broke away from the back of the house which accounts for much of the noise since it was right outside where we were standing and the oil line was right over our heads in the basement ceiling but we didn't realize it at the time. There was so much noise, yet none of us actually heard the roar like a train which many people heard nor the whine some others heard. We prayed! We were concerned that Daddy was OK so we were not entirely thinking about ourselves. As soon as the wind ceased I went upstairs. Looking out the breezeway window to the west, the cars were both still there though there were 2x4's sticking out the north side of the good car and the trailer was sitting on top of the garbage can and stacks of debris were stashed alongside both cars and the trailer. Electric lines were all down, so I was afraid to go outside. I looked out the east window of the kitchen and saw Taylor's house was gone, Bette & Bob's were standing, but the one north of their s was gone. Our big tree in the front yard got hit by something if not the wind and was broken about 7 feet above the ground. It was lying on the neighbor's car. One good thing the wind did was to take the condemned house and garage across the street from Nelsons. The new split-level south of it which had never been occupied was taken.

Directly south in line with our house is Nelsons, then Rileys, then two brand new homes. Both new homes went and their debris was what hit all the homes between them and ours. The first house looks the worst, then Nelsons have lost the south end, and ours has lost all the windows, siding, and roof shingles. We have holes in the garage and house roof from flying boards, they are repairable. The village survey states the damage at $10,000 and we were lucky! I could go on and on. The stories are frightening, sad, and laughable at times, and for some families, very bad.

Don & Sue Meyers,
146-63 1/2 Way, Fridley

(Don was the Athletic Director for Fridley Schools at the time of the tornadoes)

Sue-I was picking up hail, it was really large! I looked up at the storm and saw greenish-looking cups in the sky. Both of my grandparents had been killed by the 1924 Sanborn Wisconsin tornado and due to this family experience, I was raised to take shelter immediately if anyone of us ever saw clouds like this! I got Don and the kids down in the basement and I huddled on top of our two children and Don was covering me with his body. We could hear the terrible sound of the tornado!

Afterwards, Don had to go to the Junior and Senior High Schools to check for any damage. We got in the car and drove down Mississippi Street and drove over downed power lines that were still sparking! It was very dark without the street lights and you could hear sirens everywhere! It seemed surreal! When we got there, both of the school were hit! Don felt that the kids and I shouldn't be there so he drove us back home by way of Hwy. 100 and we saw the extensive damage done to the 100 Twin Drive-In Theatre. This all happened before the second tornado hit! The second tornado hit and followed almost the same track as the first one and did even more damage, if that were possible! Unfortunately, all communication was gone and our relatives were almost hysterical from worry about us, especially after what happened to my grandparents. The next day, the authorities issued tags for us so that we could move around our area due to all of the looting that was going on. For that year and the next, every time the skies darkened, all the neighbors would go out and watch the skies to see if any tornadoes were coming. All of us really respected storms and would go to the basement if the sirens would go off.

Don-I was the Athletic Director at the time of those tornadoes and it was amazing that no one was killed at the schools that were hit that night! The Junior/Senior High Schools were severely damaged along with Rice Creek Elementary, Parkview Elementary, and Hayes Elementary. The Fridley Historical Center building was right next to Hayes Elementary and didn't suffer hardly any damage while Hayes was completely destroyed! The two details I vividly remember from that night were at the Senior High School, there was a large football field scoreboard supported by two huge steel beams. The tornado tore off the sign and flung it across the street hitting the Fridley water tower. It left a big scrape on the side of the water tower and we knew that because the scoreboard was the same color at the scrape on the tower! It was amazing that the water tower stood up to those winds! The second detail was when I drove Sue and kids home after the first tornado, we drove by Ed Sworsky's house on the corner of 7th Street and 61st Avenue. His house suffered some siding loss and part of the roof was gone. When I drove home after the second tornado, his house was completely gone!

Pete & Mary Bonesteel, 169 Horizon Circle, Fridley

We moved to this house in 1964 and two tornadoes hit us that night! The first tornado did minimal damage to our home, mainly wind damage. My neighbor, Al Opsahl, and I walked down the street to Main and we saw a garage that had been rolled into a ditch. We helped the lady who lived there take care of some of her garage stuff from being ruined by the rain. My wife owned a beauty shop on the corner of 42nd & Central called because she heard about the tornado's damage and I tried to convince her to stay but she insisted on coming home. At this time, there was another tornado coming and it was raining with really dark clouds above. Mary came home and I met her at the garage, as soon as she pulled in, we headed for the basement. We heard the tornado coming, it hit Northern Ordinance before hitting us. It seemed that the storm used that debris to cause damage to us! This tornado destroyed our garage and the adjoining breezeway! A big 2x4 came through the front of the house blocking us from going out the front door. The strange part of the storm was that Mary had a china set sitting on the dining room table and with all the damage and the 2x4 crashing into the house, that set didn't even move on the table. No damage at all. The set belonged to Mary's grandmother from 1912 (butter dish, creamer, and sugar bowl). Mary's mom had just given it to her at Easter time because we had a new home of our own and that Mary could have them now! To this day, we respect the sirens and go to the basement whenever we hear them!

Helen Remmers,
8065 Groveland Road, Mounds View

My family and I hurried to the basement, during those days everyone went to the SW corner of the basement, by the pool table ready to hide under it if we had to! We were very fortunate and the damage to our home was minimal but the devastation in Mounds View was incredible! Especially by the Pinewood #4 area by County Road I. My husband worked for T.H. construction as an excavator and worked on many of those houses, not only before the tornado but in the aftermath. Our church on County Road I, Sunrise United Methodist Church, was also destroyed in the tornado. I was on the building committee and when I drove up and saw the damage, I cried. Our pastor told us that we lost the building but not the church. We had our services at the Bel Rae Ballroom after that until our church was rebuilt. You could smell the beer as you walked in for the church service!

I worked for a chiropractor in Spring Lake Park off of 85th and the next day when I went to work, I was stopped by the police and was told to leave my car where it was. The entire second floor of the chiropractor building was gone! I was such in shock that when I got back in my car, I got in the backseat instead of the front seat! For a second, I thought "where's the steering wheel?" The storm took our power, so we didn't have water. Our neighbor across the street had an old fashioned hand pump to get water, so that's how we managed to get water. We had a gas stove which still worked after the storm and our neighbors came over to cook their meals and I would cook up a big batch of food to share as well. We all helped each other out in this time of tragedy. You really appreciate things a lot more when they are taken away from you.

The two main items I remember that left a permanent image in my mind was that in Fridley, there were bathtubs laying in the front yards

of houses that were gone! The other was that I worked with a lot of clients from the Fridley/ Spring Lake Park areas and the children of the patients would come in and they were terrified and very "clingy" to their parents. You could see in the children's eyes that something terrible had happened.

Stan & Joan Landberg, 5144-4th Street, Columbia Heights

My wife and I were listening to WCCO and they had the warnings out about the storms and so we went outside to look. The sky was a greenish, gray color but the strange part was that it looked like cotton balls hanging down. Almost like looking from the bottom of a jar of cotton balls. It was very still outside with no wind at all. My wife and kids went to the basement and I was in the kitchen looking out towards University Avenue between the garages, about two blocks away between 51st and 52nd , and I saw a Skelly gas station sign flying down the street and I high tailed it down to the basement with my feet barely touching the stairs! My wife was 8 months pregnant with our son Paul at the time and she had a hard time fitting between the wash tubs. Our two daughters were under the wash tubs with pillows and the lights went out. I heard the kitchen window upstairs banging back and forth due to the sucking action of the tornado.

Afterwards, we went upstairs and looked at the damage. We were lucky because we lost a few shingles and the back screen door was sprung but to the west a couple of blocks, the damage was much worse. I lit a Coleman lantern because all the lights were out and my neighbor asked me how come we had lights and they didn't!

Myron & Donna Anderson,
924 Raliegh Lane, Spring Lake Park

Tornado warnings must have been made that day because I remember saying in school that I wouldn't mind seeing a tornado sometime as long as I wasn't right in it! I taught at Spring Lake Park High School at the time. After school on May 6th, I took the kids and went down to Locke Park to get some pond water and stuff for class the next day. It was real warm and muggy out! After supper, we started cutting the kids' hair. Tornado warnings came on the TV. It was soon reported that tornadoes had hit Minnetonka. It sounded as if the path would be more northeast. We took the radio down in the basement, but I told Donna there wasn't any need to be down there yet and scare the kids, etc. So we came up and I kept an eye on things. It began hailing real hard and the stones were very large, baseball-sized! I put some in the refrigerator. I also dashed out and drove the car in the garage so the windows wouldn't get broke. Approximately 20-30 minutes before the tornado hit, civil defense sirens were blowing. This probably made us more alert than anything. I was watching out the west window and saw a very black cloud advancing. It looked like a big curtain. All at once the air became dead still. I knew that was it! I said, "we better get down there...hurry up!" As we were going down to the basement with the kids, the wind upstairs became furious! The lights went out. We ran over to the SW corner and huddled down. We put our heads down and tried to huddle the kids in close. All hell broke loose! The main noise was breaking glass. All kinds of junk and glass were blowing over us in the basement. My main words were, "keep your heads down and we'll be OK...don't look!" The storm lasted longer than I'd expected. It's hard to tell but it must have blown for 10 minutes or so. As soon as I was sure it had quieted down enough, I stood up and looked out the window looking south. It looked like the pictures of bombed-out areas or similar to the tornado damage in Michigan. It wasn't long before we smelled gas and heard it hissing. This scared me plenty. We knew we'd better clear out. We managed to get out by the garage door after pushing things to the side. We were scared of downed wires. Our neighbor was running down the street screaming about not finding her baby. My next door neighbor, Dale Davis, and I ran to the house to look for the baby. We heard the baby crying from under a wall that had been set by the tornado on top of the crib without crushing it! We lifted off the wall and the baby came out of it miraculously without a scratch!

Our house was completely demolished and our neighbor to the east came over and invited us to stay over there for awhile. Later, I came back to our house to put a few things in the car like clothes, encyclopedias, family pictures, etc. We were thankful to be alive! I envisioned a bigger loss than we actually had. We tried to get to a telephone and call our relatives to let them know we were OK but couldn't. Somehow, our old Plymouth didn't get too damaged and I managed to get the car to the intersection of Highways 10 and 65. I got two flats on the way there and the filling station couldn't fix them. They didn't have electricity. I was able to get the tires finally inflated and get to a working phone in which I was able to let the relatives know we were OK. It's amazing how much stuff was actually saved but it was a quite a clean-up....fiberglass, wood, and crap all over the place...broken legs on the sectional, smashed up bedroom sets, etc. I was lucky to get lots of help. Ted came. Mrs. Murray helped. Several teachers helped. We moved all the stuff over to Fletcher's garage. The next day we got our house on 10425 President Drive. That was a lucky break! My sister Della came down and helped us move in. We got around $132 from the teachers at work. That was a lot of money back in those days. We bought a rug and a vacuum cleaner with the money. A lot of people helped and it was good experience in human relations. People are quite compassionate in times of turmoil except the looters.

Donna-I was cutting our boys' hair in front of the TV set when the warnings came on saying that the tornado was coming from Lake Minnetonka. I told Myron that we're heading for the basement! It looked like a black curtain in our front window. So, I grabbed our 5 kids (ages 7 to 16 months) and ran down the basement with Myron right behind us. We went to the SW corner of the basement and the lights went out. One of our sons tried to run away from us and I grabbed him back into the corner. Afterwards, we had to push the basement door open due to all the debris in front of it. The house was in really bad shape with only a few walls still standing. Some of the strange things that the tornado did was that our kitchen clock was still hanging on the wall stopped at 7:09. Our cookie jar, a pig dressed up with big red buttons, had all the buttons plucked off but the jar was still there in good shape!

LouVerne & Della Moen,
7717 Able Street, Spring Lake Park

It was my birthday (Della) and we never got a chance to eat the cake and ice cream due to the storm! Our daughter, LouAnn, was just coming back from the store when the sirens started to sound. The people at the store told her to go home right away because a tornado was coming. All of us went to the basement and I was the last one going down there, I was sort of adventuresome! We went to the couch in the NE corner of the house and prayed. I believe that God answered our prayers and no one was severely injured or killed! I went to the SW corner, holding

Steve (2 yr. old), and tried to look out the window. I never really saw anything threatening and was wondering what everyone was talking about. And then the basement window blew in on top of Steve and I scratching us from the broken glass. Then I knew it was serious! The thing I will always remember is the suddenness of the storm and that you never think that it is going to happen to you!

Our first reaction after the storm is what are we going to do? It's strange because our car was in really good shape with no apparent damage but there was a wall slid underneath the car! And the wall wasn't from our house or our neighbor's!

Vern Moen,
7717 Able Street, Spring Lake Park

At the time, I was 12 years old. It was my mother's birthday and we were getting ready to celebrate and eat supper. LouAnn, my older sister, was sent to the corner store for some milk and bread. We were wondering if she should go because the sky was really dark and green. When she got back, we sat down to eat when it started to rain and hail. My dad got all of us downstairs to have eaten the cake and ice cream down there and then the lights went out and the big hail started coming down. It sounded like someone was upstairs and the hail was huge-softball sized! My dad went upstairs and came running back down because it started to break the windows making smashing/crashing sounds. He told us to hide. My mom was trying to look out the basement window when a 4x4 came flying through the window almost hitting her! My brother Norm and I hid in the SW corner of the basement and we heard the tornado coming! It sounded like a freight train but louder. My brother was protecting me from the top when we started coming up off the floor and we saw the floor boards above us pealing up! Then, it was over!

It was raining on all of us and my dad got us all back together as we huddled by each other. We went upstairs and couldn't believe the damage. The whole south end of our home was gone (3 bedrooms and bathroom). We sat in the garage which was still standing and we saw that the tornado had taken one of our walls and slid it under our car, a 1958 Buick! It seemed that the tornado simply lifted the car and slid the wall underneath and set the car back on top of it. About this time, my uncle Jim came over. He had to park on the corner of Able & Osborne because of all the debris in the way and then walked up the street to our house. The tornado had blown a tree through his bedroom wall and so he was coming over to stay at our house but our house was in worse shape than his! We walked back to his car and we went to stay at his house. As he drove over the downed power lines, we were told not to touch the door handles. I'm finally getting over my fear of storms after all these years, it was a night I'll never forget. Whenever a storm comes, I make sure my family goes to the basement for safety.

LouAnn (Moen) Gulbranson,
7717 Able Street, Spring Lake Park

I was sent to the store to pick up a few things for supper and it looked fine when I went there. On the way back the sky was a pea-green color and very humid/muggy. My family and I ate supper and afterwards the sirens went off. We went downstairs with the cake and ice cream because it was my mother's birthday. Our dog, Tula, was the first one down in the basement and she never went down there! I heard the screen door upstairs flapping so I ran back up to shut it. When I looked out the screen door, I saw the tornado coming down over our neighbor's house across the street. It was a big, black/gray spout and I knew it was going to hit us! I ran down screaming to the basement! The windows of the basement exploded and my dad flipped the couch on top of my brothers & sister. A big, gray pipe and a 4x4 piece of wood came flying through the broken window and almost hitting us. The sound wasn't really like a freight train but more like a high, wind sound! Then, it was over!

The next day we came back and it was my birthday (#13). My dad and I sat on the front steps and just bawled because we lost everything. We were lucky though, we had each other and no one was injured or killed!

Carol (Lessard) Butgusaim, Fridley Terrace, Fridley

I was 18 years old at the time visiting a friend who lived in that trailer park. There were four of us who were in the trailer when the sky turned a very strange, greenish color. It was absolutely still and every leaf on the trees didn't move! The radio was on and stated that there was a tornado coming (first one)! We laid on the floor underneath the kitchen table when it hit! I will never forget the noise! It was like a deafening roar, extremely loud! The trailer was turned over on its side. We were young and you never think that anything could ever happen to you but when we crawled out and saw that total devastation, we were in shock! People were wandering around in a daze without a purpose. Some were giving each other hugs, thankful to be alive. We went around helping people get out of the debris or trapped underneath . Cars were flipped over and it looked like a bomb went off! My friend's car was wrecked so I ended up walking home that night to my home in Columbia Heights. Years later, when I think about it, I wonder how I or anyone could have survived that tornado.

Bob & Delores Peters,
101 Ventura Avenue, Fridley

We weren't home when the first tornado came through.My wife was shopping at Apache Plaza for a Father's Day gift and our children (4) were with me. When we came home, all of our windows were out and we hurried up and boarded them up to keep everything in our house dry. Our house was across the street from the Midland Oil Company.

Then the second tornado came! We heard it and saw a big black cloud on the ground with all kinds of debris whipping around. I got all the family in the basement and sat everyone on the couch with my wife and I covering the kids with our bodies. It sounded more like a threshing machine than a freight train. When the tornado took our house, cement blocks and the pieces hit my wife and I. One of the blocks hit me in the head and I spent 2 hours at Mercy later that night having the doctors clean me up. The blocks were lower quality cement due to all the sand in it, there was a lot of sand in my head wound.

The tornado hit Midland first to our west and peppered us with melted tar from its roof! When we all got cleaned up, we found that the tar had gone through all our clothes. We had to use diesel fuel to wash our hair to get all that tar out! All the houses on our block were destroyed. We had lived in our house since November of 1958 and we never did live in that house again after the storm. The looters that came around really upset us because of the tragedy that struck us and all of our neighbors and here these people came to take advantage of the situation! The next day, we came back to salvage what we could. Later on, people in Grantsburg, Wisconsin had found some of our cancelled checks!The tornado had carried our papers/documents all of that distance!

**Vern Stegeman,
National Tea Grocery Store,
University & Mississippi, Fridley**

I was the assistant manager of National Tea when the tornado hit our store. It was about 7:30 p.m. when it hit. The sky was very dark and electricity flickered before it went out and we didn't know what was going on. There was piped-in music but no news until a guy came rushing in telling us that there was a tornado coming. Just then, the lights went out and we saw the National Tea sign blow off and the Standard Oil sign across the street blow off and hit the fire station across the street. I told everyone that where we were standing (in the front of the store by all the windows) was not a good place to be and that we should all go to the back of the store. As soon as I said that, the far end window blew in. We rushed to the back of the store and took shelter in the meat room. The noise sounded like a freight train and the front end of the store came crashing down! It all happened so fast and you didn't have time to think about anything, you just did what you had to do!

Afterwards we all went outside and the damage was incredible! The windshield was blown out and the roof was damaged on my new car, a 1965 Plymouth station wagon. I wanted to go home but all the streets were blocked off and the police wouldn't let any vehicles go down the streets with all the damage so I ended up walking home. There was damage everywhere and I was worried about my family and what condition my house was going to be in. When I got home, the only damage was to the paint and some shingles missing. The total cost of damages was $800. But, many homes in my area were completely destroyed, we were very lucky. After the storm, a WCCO van came and parked in our grocery store lot and interviewed me and some others that had gone through the storm. I will never forget that night!

Bob Baker,
644 Ballantyne Lane, Spring Lake Park

I was 8 years old when those tornadoes came through and we were about a block away from the total devastation. A lady was killed about 10 blocks away on our street. A 2x4 was

shoved through her as she was going to her basement. Our relatives were extremely concerned when they heard a woman was killed on Ballantyne Lane in Spring Lake Park. Back then, there were no cell phones and the phones lines were down so no one could let relatives know if they were OK or not. About 100 yards to the east of us, houses were flattened. Friends of ours, the Hornseth family, their house was completely destroyed except for the bathroom on the main floor, that was only a few blocks away. We lost our fences and roofing but the main structure stood.

It was very hot and uncomfortable that day. When the skies became black as night and the sirens started going off, it didn't surprise me at all. We went down to the basement and my dad lifted me up to look out the basement window to see the tornadoes hovering/going over our house. You could hear the shingles being ripped off the roof but the noise I'll never forget was the moaning and screaming of the ductwork in our house! The tremendous drop in barometric pressure caused the ductwork to contract/expand and the noise was unbelievable! My mom had brought down fudge, candy, and pop for us kids and that was a rare event. The lights went out and it seemed more like a camping trip to me than anything else.

The next day, we came out and surveyed the damage and walked around the neighborhood. My parents felt bad about taking pictures of all the devastation. These people were our friends and neighbors. The National Guard had set up checkpoints to make sure you lived in that area and to keep out looters. There were a lot of downed power lines so you had to be careful when you walked around the neighborhood. All the new housing was split levels because that was the popular style of houses in the 1960's. You can still tell today what areas were destroyed by the tornadoes by the style of homes in that area. Afterwards, when storms would come, all the neighbors would walk to the end of their driveways and look for tornadoes.

Many schools were destroyed that night and I attended Kenneth Hall Elementary School which was one of the schools that was hit by the tornadoes. I finished the school year in one of the Blaine elementary schools for half days while the kids who regularly attended that school also went half days.

Melvin & Lou Persson,
7871 Jackson Street NE
Spring Lake Park

The tornado hit about a half hour after I got home from work. I drove a delivery truck for Ran Meat and my route that day was the western side of the metro in the Buffalo/Excelsior area. When we heard the warnings coming in, it seemed that the storm was coming from where I had just come from. Willy Strom, my next door neighbor, and I were watching the darkening skies from our front steps when we saw the tornado coming! I ran in the house and went in the basement where Lou and the kids were seeking shelter in the well pit. We heard the sirens going off and the lights going out. I forgot my guns that were in our bedroom, so I ran upstairs to get them. By the time I got back down to the bottom steps in the basement, the tornado hit. We heard windows breaking and debris hitting the house. Afterwards, we went upstairs and the mess was unbelievable. Lou found slivers of wood sticking out of the siding. We were lucky, our windows, siding, and roof were damaged but the house still stood. On either street to the west and east of us, the houses were completely destroyed! The tornado had passed over us and came down and hit Spring Lake Park High School several blocks behind us. We had a white picket fence that took me years to put up that was destroyed. The wind broke the fence into sections and spun the sections of fence around the post in the ground like a top! The damage to our house was $3400 and we had paid $10,500 for it in 1960!

Herb Saxon, Minneapolis

I was working as a trucker and had a pickup at the Sears store in Minneapolis. The dockhands started hollering that there was a tornado coming, so I took off for Spring Lake Park where my parents lived. I had to drive over power lines to get there. My cousin and his wife lived in Fridley so I went over there after the tornado came through. The only way to get around there was by riding bicycle. We rode our bikes and there was trash everywhere. There were only downed branches and a power outage in Blaine. A friend of mine, who also was a trucker, worked at the Midland company in Columbia Heights which was heavily damaged by the tornado. He was parked at the dock unloading and the roof came down on his semi and crushed the cab! Luckily he was inside unloading his freight and not in the cab.

Mary Saxon, Elk River

All I remember was that my dad worked for the Ag/Stabilization Service and had a special pass to get into the roped off areas damaged by the tornadoes. He took me out there two days after the storm and I saw a tree that was pierced by a corn stalk! We saw a dead cow that had a long piece of wheat stuck right through its stomach and a barn that looked like Indians had attacked it with arrows, except it was wheat! I was only three years old at the time but I will never forget what I saw from those tornadoes!

Geoff Olinyk, Fridley Junior High School

I was in 10th grade, 15 years old, and attending a Boy Scout Council the night that the Fridley tornado hit! My parents and I didn't hear the sirens but felt a rumble. The council was being held in the gymnasium with at least a couple hundred people in attendance, my dad being a scout leader, when the roof came off! It looked like a big fist crumbling a piece of aluminum foil being lifted straight up. The first few courses of cinder block came tumbling inward toward the people sitting in the bleachers and everyone scrambled downward toward the main floor. The lights went out and it was pandemonium! It was amazing that no one was killed! Afterwards, we went outside and it looked like an atomic bomb went off! There was sheet metal and debris all over with cars rolled over. We had a terrible time getting home that night due to all the road closures and debris blocking streets. We saw a bathtub with a quilt in a tree about 6 or 8 blocks away by Silver Lake Road. The 100 Twin Drive-In Theatre was destroyed with all the speaker poles bent to the ground. It was a night I will never forget!

Robert Zimmerman,
Route 1, Osseo

I was 13 years old when the Fridley tornado came through by our farm. The tornado hit Evelyn Stezler's parents' farmstead which was located where the Fleet Farm is today on Hwy. 169 and 85th. The barn and shed were completely destroyed and one of the wood support beams was hurled into their house which was at least 100 feet away. That beam measured 8x10 inches by about 60 feet long! Evelyn later used part of that beam to build her new home. My parents and I were watching from about a mile away and the funnel looked a white gray color. We seldom went to the basement but we did and my dad watched the storm from above. Afterwards, I saw lots of uprooted trees. It seemed that the storm went east along 85th due to all of the uprooted trees in that path.

Carol (Scherer) Christenson, 6415 Jackson St. NE, Fridley

I was very busy with 4 little children the night of the '65 tornadoes. My husband was in Minneapolis for a meeting and he called at about 6:30 p.m. informing me that a tornado had been sighted in the Lake Minnetonka area but not to worry because there was no way it would ever hit us! About 7 p.m. the sky turned green and we headed for the basement. I got all the kids in the SW corner but forgot the flashlight, I told the children to stay put and I would be right back. When I went upstairs, it was pitch black outside. I ran down the stairs and it hit! We had very tight windows in the basement, the house was almost new but sand came blowing through around the seals! The noise was unbelievable and our children were really upset.

After the tornado hit, I went upstairs and surveyed the damage. We lost some windows and the tornado moved our garage off the foundation. The garage door of our detached garage was wrapped around our neighbor's tree! Two or three houses down from where we were, the houses were gone! Lorrayne Weiss came over and told us to come over to her house. She knew that my husband wasn't home and that we could be together over there. During this time, my husband, Gordon, left his meeting when he heard that Fridley had been hit! It took him four hours to get home due to all the damage and debris in the roads.

I was really busy with our children but a few things still stick out in my mind. A house down the block was completely destroyed but at the time of the tornado, the people who lived there had gone to McDonald's and in their hurry, left the bag on the kitchen table which was the only thing not taken by the tornado! It was so quiet and still before the storm hit. You just felt that something bad was going to happen, it was in the air! The entire night was a nightmare! The next day, we went out and surveyed the damage and it was total devastation!

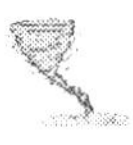

Bernadine (Scherer) Matushak, 1666-29th Avenue NW, New Brighton

We had nine girl scouts at our house that night for a meeting. We stood and watched the storms and saw the big tornado that hit Fridley and Mounds View at around 8:30 p.m. One of the girls exclaimed, "What is that?" As the tornado hit Fridley to the northwest, it was completely illuminated by lightning. Every time the lightning flashed, you could see the funnel. Then, it disappeared for a short time and came down again as a monster tornado to the north of us in the Mounds View area!

One of the parents of the girls had been hit in Fridley and called to say they were picking up their daughter early. We all went inside and watched everything happen on T.V. Back in those days, it was hard to watch because it was so dark outside and the cameras weren't that good in the dark. We were lucky, it didn't even come close to us but we could see the tornadoes and the devastation they caused!

Linda (Nagle) Johnson,
8801-60 1/2 Avenue North, New Hope

I was only 3 years old at the time but I'll never forget that day! The sky was very yellow and had a greenish tint to it! Very unusual! My dad was watching the sky and it seems like all dads try to be tough like that! We hurried down to the basement and my mom flipped over the couch and hid us under it to protect us from the tornado. I heard the wind and was scared but fortunately the storm went by us and my dad saw it moving towards Fridley/Spring Lake Park.

I've been to the historical society in St. Paul and they have a display over there and it was exactly the same as what I went through in the storm! They actually have a display where you go into a room and is set up to look like a basement from that time. They have WCCO on the radio and they have the winds going and have a great slide show. Go to it, I really enjoyed it!

Debbie (Kaufmann) Hensel,
3539-2 1/2 Street, NE Minneapolis

I was 12 years old when the Fridley tornadoes hit. I lived in Northeast Minneapolis and

my home was fine but I remember the dark, green ominous skies and feeling very scared! The next day, I went with my grandmother to check out the rental home she owned in Spring Lake Park. We had trouble getting to it because of debris everywhere and the fact that the police had roadblocks on the streets that were open and were only letting residents of the area in. My grandmother had to convince them that she needed to get into the neighborhood to check on her rental home. When we arrived at the home, we found damage but the house was still standing. One of the biggest problems was a lot of water in the home.

A few years later, when I was in junior high school, my home economics teacher shared with us her family's experience in the '65 tornadoes. She lived in the Spring Lake Park/Fridley area. She showed us an aerial photograph of her neighborhood. Although damaged, her home was still standing, but every house around hers was not. Her family opened their home to the neighbors whose homes were destroyed. They received an award from the Red Cross for their generosity.

Jan Davis,
8503 Davenport Street NE, Blaine

We were renting a twin-double bungalow on a block off of Central and 85th Avenue in Blaine. We had no basement and I laid myself over our 1 1/2 year old son as the strong winds and pressure came pushing on the walls. It popped the windows and sounded like a freight train! Just as I thought the wall was going to cave in on top of us, the pressure let up. The roof leaked water into the house. My husband was at a Army Reserve meeting at the Fort Snelling area and he had a lot of trouble getting home because of all the debris blocking all the streets in the area. All the street lights were off and the police were barricading the streets. He thought for sure our bungalow would be gone! After it went through the first time, neighbors started contacting each other. Us without a basement went to a neighbor with a basement, so when the tornado came through again, my son and I were safe in our neighbor's basement!

Gayle Kirkham, 501-67th Avenue, Fridley

I was almost 10 years old when the Fridley tornadoes hit in May of 1965. I had been at a birthday party that afternoon at Lake Calhoun. When we arrived home around 5:30 pm, we heard there were storm warnings out. I hurried into the house, had some supper and as we were eating, the sirens started sounding. We went to the basement, into my dad's workshop and huddled together in the corner. My mom, brother, his friend, and I were the only ones home. We were supposed to be at the Junior High School for a science fair, but decided to stay home when we heard the storm warnings. My dad was working late at Honeywell in Arden Hills.

As we were huddled in the basement, we heard a very loud roar that sounded like a freight train going right over the top of us. Our ears hurt from the pressure of the storm. We heard glass breaking, but had no idea what was happening. Because there were six tornadoes that hit our area that night, we stayed in the basement for quite a while, as warning after warning kept coming over our transistor radio. Because of the sound and not knowing what was happening, we were very scared. Between storms, we tried to call my brother's friend's family to let them know he was safe, but there was no phone service! There was also no electricity.

My dad got off work around 8 pm, but couldn't get through the streets and neighborhood to get to our house for quite awhile. When he got home, we heard what the area looked like around our house. After he came home, things started to settle down a bit and we went upstairs to see what had happened.

Our front window had blown out and glass was lodged in the wall. The windows had blown out on the side of the house, in the kitchen as well as the back door. Debris was everywhere and most of it wasn't ours! We found a swing set in our backyard wrapped around our shrubs! Our house wasn't in too bad of shape. Then we looked around.

Next door, the garage, right outside my bedroom window, was gone! It got worse from there! The house across the street had the roof lifted up and the living room drapes were hanging out over the wall, with the roof back down on top of them! The houses down the street were damaged more and more until you got to the last house. The woman that lived there had gone back upstairs to open a window when the

tornado hit and she was blown through the wall. She was found in the backyard injured badly. The neighborhood men found a door that had been blown off its hinges and used that to carry her from the backyard to the street to await an ambulance. Most of our neighborhood was totally destroyed.

The National Guard came into the area to guard against looting. They also helped with some cleanup. The electric and phones were out for many weeks. We had a gas stove and oven, so my mom stayed in the house and cooked up food for everyone. Neighbors would bring over their food from their freezers and she would cook it up so they wouldn't have to throw everything away.

The neighborhood pulled together and everyone helped each other clean up. Our church was used as a Salvation Army food and clothing shelter. It was a very scary experience and that fear lasted for many years! Our elementary school, Hayes, was demolished, so school was ovcr for that remaining school year! The following fall, we had to do split shifts for school until they could rebuild the schools that were damaged from the tornadoes! It took a long time for everything to be rebuilt and totally cleaned up.

Jerry Stegeman, 6350 Van Buren Street, Fridley

I had just turned 14 years old just days before May 6, 1965. Some friends, my two brothers and I were playing baseball in our street. It was after supper, about 6:30 pm, when the sky turned very black and the wind picked up. Soon the sky turned a scary, greenish color that none of us had ever seen before. Rain began to fall so some of us went home, some went to Palmer's garage to escape the rain.

My brother Tom and I went home. Our two sisters were there with us. My parents were at work. My mom worked at the Foshay Tower. My dad was an assistant manager of the National Tea grocery store off University Avenue. My middle brother, Jim, stayed at Palmer's house.

At our house, the four of us went to the basement after Mom called to warn us of a bad storm. We huddled under a full size pool table in the family room. Everything became very still outside until we heard a terrible roar! We heard hail hitting the house and windows breaking. Our ears were affected by the pressure. Soon, all was quiet.

At work, my father saved several lives by getting people away from the large front windows. He'd just gotten the last person away, when the windows blew in! He barely gotten to safety himself. Everyone headed for the dairy cooler where they found safety. When they emerged after the storm passed, the entire front of the store had caved in. Police and fireman from across the street came and helped them to safety. One of the large light poles had crashed our station wagon. Both parents had trouble getting home that night.

At the Palmer house, Mrs. Palmer had barely gotten the boys inside when the garage was picked up and blown away. From their house on Jackson Street and west for miles, the storm totally leveled blocks of homes, often just leaving a bathroom wall and toilet on the 1st level up.

I remember walking through the neighborhood the next day. We saw 2x4's driven through walls and trees. Houses by the dozens were leveled to the basement. Trees were twisted and shredded; debris was everywhere. Then, we heard people were killed and many were injured. We were lucky, we only had a few broken windows. There was a science fair at the Junior High School that night and though the storm demolished a lot of the building, there were no fatalities at that school. My aunt watched it all transpire from her Hilltop home!

Jan & Ralph Reiter, 1130 Wyldewood Lane, Spring Lake Park

We lived in a four plex right off of Hwy 65 and 2 blocks south of Hwy 10, across from Spring Lake Park High School. It was about 7:00 pm and we were watching TV (Davey Crockett) when they blurted in with a tornado warning! Of course, like any other time you think it is not going to hit your place, we weren't real cautious. Then it got real humid (it was a hot day) and it started to hail. All of a sudden, it got real still and Ralph went outside to Hwy 65 (about 400 yards away) to see what was going on. He told me to open all the windows a couple of inches to equalize the pressure if a tornado came. So I did. It got real dark and he looked to the SW where it was even darker and he could see things being tossed. There was a roar like a large train running over you! He came into the fourplex and made all of us get into the SW corner of the basement apartment.

All of a sudden, our ears were popping like when you are on an airplane. We heard things flying around upstairs! Then, something big hit the upstairs. We heard it but had no idea what it was. Later on we found out if was the neighbor's roof that made a hole in our bedroom upstairs. You could feel the building just shake! We were very scared at the time and for a whole year after that. Every time we heard there was a tornado warning, we were terrified! We think the tornado lasted probably 2-3 minutes but to us it seemed like an eternity! You are so scared we wanted it to end!

After the tornado struck, it was light for a little time but then it got dark again. It started to rain. We heard there was a second tornado but we never saw it. We had a portable radio so that we could hear what was happening. They told us to stay in the house because of all the downed power lines, yet there were people outside looting. We heard one say, "Oh, I found a radio", and another, "I found a purse". We just wanted to get those people. Here we were having a tragedy going on and they were stealing. It just didn't seem right. There were no telephones because all the lines were down. We couldn't let our relatives in St. Cloud know what had happened and that we were okay. They were worried until we were able to call them the next day when we got a pay phone that worked.

After it was all over, we went upstairs into our own apartment and we had a big hole in our bedroom, debris all over the place. Our TV had flown across the room and generally it was just a big mess! That night we picked up as much as we could with a flashlight. We put plastic over the big hole so at least we could get some sleep. The next day we could really clean up. When we got up the next day, we saw the full extent of all the damage. Across the highway, the high school's roof was gone. The realty business was totally gone. Several houses were gone or half there. Debris all over the place! Branches, trees, power lines down. We saw a Volkswagen wrapped around a telephone pole! A lot of businesses were wiped out. The National Guard was out the next day to protect the homes from looters. You had to show your ID that proved you lived in the area to get in. Our car radiator had a hole in it from wood slivers and debris so we couldn't drive too far. We carried water with us to fill the radiator when it got too hot. It took about a week before we got our place straightened out and fixed.

Rochelle & Dale Hartje,
3027-6th Avenue North, Anoka

Dale and I were at:

- Target in Crystal with our three kids... ages 2, 1, & 0... fourth born in 1968.. not even a twinkle in our eyes!
- The store cleared out because of the tornado sirens!
- We couldn't figure out where all the people had gone.
- We didn't hear the warnings!
- We were the last ones in the store, except for the employees!
- They told us to get home quickly!
- We lived in Anoka.
- We raced to our car!
- We drove the distance from Crystal... through Osseo.
- Numerous huge trees were down, on Jefferson Highway in Osseo.
- It was extremely windy..making it difficult to keep the car on the road!

- No seat belts or car seats in those days.
- We arrived home safely!
- We went directly to the basement!

Later, it was a shock, to hear the news about all of the damage caused by the tornadoes!

Tim Loberg,
5950 Camden Ave. N.
Brooklyn Center

I was 10 years old when the tornado came through. Our Little League game was cancelled due to the bad weather and I got a ride to my friend's house, Bruce Ranking. My dad had to stay because he was the scorekeeper. When we got to Bruce's house, it was raining, cold rain with hail and the sky was an icky green/black color. You could feel the awesome pressure in the air. Bruce's sisters were in the basement and kept coming outside screaming at us to come in and get out of the storm.

About a half hour later, Mr. Ranking drove me home and it was dead calm. My dad watched the storm coming and we saw a spiral-like cloud come down and take the roof off of Kevin Nitzell's house across the street! At that point, my dad said, "Time to get in the basement!" The

tornado lifted up again and came down over the Mississippi River and it nailed FMC! We came back out and it was a cold rain again and the sky looked yellow/green. The only house in our neighborhood that suffered damage was the Nitzell home.

Denny Brown,
Maguire's Restaurant, Arden Hills

I was working as a busboy at the restaurant in my senior year when the storm came. Someone came in and said there was a tornado coming! We got all the patrons to back away from the windows and some of us went out the back door to take a look. We heard the sirens beginning to go off all around us and to the west, the sky took on a ugly, green color. It was hot and sticky and you could see the debris/junk flying around the funnel. I was far enough away so that it didn't seem so bad. About a month later, I went over to Lake Minnetonka and there was still lots of damage from those tornadoes!

Jim Smith,
11551 Larch St., Coon Rapids

I grew up in the Thompson Park neighborhood in Coon Rapids. My parents had bought that house in 1954 when I was born. In 1965, during the demolishing storms, I remember most of the people in our neighborhood outside looking for tornadoes and everyone had a radio set on channel WCCO AM because we didn't have FM at that time. Everybody was getting a "blow by blow" account of what was going on with the twisters in Fridley and Spring Lake Park. There were so many people outside with portable radios that it sounded like a big stereo! Some people were standing on the roofs watching the strangest looking clouds I had ever seen! The clouds were like a camouflage color, mixing, swirling, and churning altogether. The sirens were going off and you could tell the people were really scared. Up on the corner of Hwy. 242 and Hanson, the pine trees were destroyed in different ways. On one corner, the trees were simply twisted out of the ground. The opposite corner, all the trees were blown straight down.

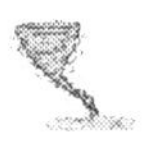

Mark O'Neill, 861 Rice Creek Terrace, Fridley

I'll never forget that day! It was very humid and warm, in the 90's with a strong SW wind that blew all day long. I was 12 years old and was in Mr. Erickson's 6th grade class at Hayes Elementary School. We were going on a field trip the next morning to Chicago on the train. Hayes was destroyed that night including multiple hits on my immediate neighborhood. My Dad taught at the high school, which lost its roof, the junior high was damaged along with Spring Lake Park High School. We had one elderly lady on the next block who was killed.

The Fridley Terrace trailer court took a direct hit where another fatality took place, a young child I believe. The first tornado hit at 7:11 p.m. because that is when the power went off in our area! We were lucky, on either side of us (2 houses), every house was destroyed or heavily damaged. There was a house down the block from us where the tornado hit it right down the middle leaving a "V" shaped cut, the outside edges intact but nothing in between!

Before the tornado hit, my dad and us kids were watching from our front door when an object flew by us like a flash and my dad said, "Ok, time to head for the basement!" There was only one siren in Fridley, at the fire station, on the corner of University and Mississippi and it was blaring! While we were in the basement, eight of us, I was really scared and you could feel the tremendous pressure as the tornado hit! After the tornado hit the area, the Prindle family stayed with us for awhile because their house was completely destroyed. I remember hearing a story about two boys, Dean Brenny and Mark Rasmussen, were riding their bikes in front of Hayes and as the tornado approached, the janitor ran out and grabbed the boys. He took them quickly to the shelter in the school. If he hadn't done that, those boys would have been killed for sure!

Jackie (Keith) Sandusky, Wayzata High School

(currently the Junior High School)

I was at the school play that Friday night and I was in the 10th grade. My role was the makeup person for all the actors/actresses in the play. It was very hot, humid, and muggy and all of us went outside to cool down and it was really tough keeping all the makeup from smearing, etc. All of us were sitting down on the south end of the building and I will never forget the yellow, hazy sky. The air felt very heavy and oppressive! It became more dark and windy with the sky turning more green and gray.

Just then, the police were driving up to us and they told us that a tornado was coming across Lake Minnetonka and that we should get to shelter inside the school. All of us went to a small, narrow hallway in the back of the gym and sat down on the floor. We were so far back in the school, you couldn't hear any noise from outside. Finally, we were allowed to go back outside but only our parents could pick us up. My mother picked me up and we drove to check on some friends that lived in Navarre. This area was really hit bad and had a lot of damage. There was a lot of debris on the roads. We heard later on that one of the twisters came over our school and hit the track leaving some damage on it!

Marty Lueders,
6831 Oakley St. NE, Fridley

At the time, I was working on a government project with research balloons and air quality at the Anoka airport. I would send up the balloons at sunrise, in this case, about 6 a.m. the morning of May 6th, 1965. I remember that morning well because we had a long squall line come through at about 7 a.m. just after we sent up the balloons. My wife, Carol, was due with our second child any day so I didn't want to be too far from home and she told me to go ahead and not worry about it. So, my partner and I left for the New Richland, Wisconsin area where the balloons landed. It was about 10:30 a.m. when I got the radio message that my wife was going into labor. At noon, we had a big storm over us with lots of rain, hail, and wind. I had to get back to the Cities and used the recovery truck to drive back and arrived at about 3 p.m. Our three-year-old son, Jeff, was staying next door at the Soderstrom's and after checking to see if he was OK, I left for Methodist hospital where Carol was. In the meantime, our daughter, Kristan, was born at about 1 p.m. so I was late anyway.

About 6:30 p.m., the doctor came in because of the sirens and tornado warnings stating a tornado was coming from the Lake Minnetonka area and I decided to leave and get Jeff from our neighbors. I drove through a ton of rain and wind to get home. When I got there about 8:30 p.m., our house was completely destroyed. I had to park on Hwy. 65 and we had lived about a block to the west. The last tornado came through and tore the house off our foundation, flipped it over, and threw the floor a block away blocking traffic on Hwy. 65! In the nursery, we had a chest of drawers with all the baby stuff and we found that in the backyard, upright and none of the drawers open with nothing missing!

I checked on Jeff and the Russ Soderstrom home had all the windows gone with part of the roof missing. Thank goodness Jeff and everyone was fine! There were 17 people in their basement, a lot of people stopped on Hwy. 65 and ran to their house to go in the basement when the tornado hit. Jeff heard the sirens and said, "the fire is coming!" Russ' pickup was picked up and tossed over to the KC building across the highway! My '52 Chev, which was parked in our driveway, had the trunk lid flipped and bent over the back window. I was told that this happened after the first tornado came through. Everything in our basement was sucked out except the washer, dryer, and deep freeze!

I often wonder what would have happened if Carol didn't go into labor early and she and Jeff would have been home when the tornadoes hit. We were so fortunate that no one was injured or worse. After the tornadoes, I received a call from a person in Wisconsin stating that they had found my Target credit card along with some other correspondence!

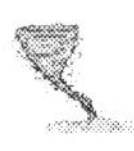

Terry Sworsky, 491-61st Avenue NE, Fridley

We played baseball that day and we beat St. Paul Park (ranked 3rd in state). I was a senior and we were having a great season and hadn't lost a game yet. After the game, we all showered up and went home. The clouds were circling above us, puffy, dark, and very green. It began to hail, golf-ball sized, we went in the house and came back out with garbage can lids to cover our heads while we picked up some of the hailstones. About a block or two south of us, transformers were blowing up and garage roofs were folding up like books being upturned and slammed shut. We ran to the basement under the steps and our ears were popping! I looked out the window and saw the telephone pole out front being snapped in half!

After the storm passed, we went upstairs and the house to our north had an upturned car laying on its side about 3 feet in front of their picture window. This really looked strange to me. The tornado had torn off our garage and porch but not much damage to the main house,

we were lucky. Just then, a man was driving down our street over downed power lines and got stuck on top of them in front of our house. He looked really nervous and we told him not to touch anything because we didn't know if the wires were hot or not. It took about 15 minutes before we were able to get him out safely, we used a big plank for him to get out with.

My dad left to go survey the damage and managed to get 4 flat tires from all the debris in the streets and our mom was working at the time. We decided to eat the chicken and rice dish my mom had left for us and we were talking about how funny/weird it would be if another tornado hit. Just then, the wind picked up and it got dark out again. There was a 2-3 inch diameter maple tree in our front yard that was blowing in the wind and suddenly went flat to the ground! I said, "Oh-oh!" We saw two high school girls walking down the street and we yelled to them to come in our house and get in the basement. As soon as we got in the basement, the whole house began to shake and creak, loosening dust from above us! This tornado was much louder, we could hear glass breaking, everything was breaking! Our roof was lifted up, set back down, and left a big crack around the perimeter of the ceiling.

We looked out the back door and saw that the neighbor to our north now lost their entire house, except for a portion of the interior bathroom wall. The upturned car was still there. We couldn't go out the back door because the eve of the roof now sagged over the door opening. I went around the front and saw the tornado off in the distance, every time the lightning flashed. It was real eerie looking!

In the aftermath of those tornadoes, I remember how loud the tornadoes were, especially the second one and how totally helpless you felt as it passed over. Our baseball team wound up losing most of the remaining games of the season, simply lost our focus due to the storms. In the ball field across the street from us, we would find lots of money laying in the mud and dirt, for a long time after. There were 20's, 50's, and 100's. Apparently, people had stashes of money that the tornadoes spewed all over. Whenever mom would make that same chicken/rice dish, we would tell her not to make the "tornado dish" anymore!

Jerry & Marilyn Manley, 660 Marigold Terrace, Fridley

I got home from work and the TV warnings were on 'CCO. It was about 6 p.m. The tornado was getting closer, Brooklyn Center, Brooklyn Park...Marilyn and our children went to the basement except our son, Mark and I. Our kitchen window faced west, Mark (10 years old) and I looked out and saw pitch black! This seemed really strange because it wasn't that late and then Mark said, "Hey dad, there's stuff flying around!" I realized then that what we were looking at was the tornado! We were so close that you couldn't see the classic funnel shape! We leaped for the basement and jumped downstairs with the others hiding under the work bench in the SW corner. The lights went out and we heard booming and crashing noises upstairs. Besides the freight train sound, the noise was a tremendous roar!

We had moved in our home in 1955 and with 6 kids in our family, we were in the midst of remodeling by adding a few more bedrooms. All the windows, back door, and roof on the south side of the house were blown out and destroyed. We were lucky, most of the worst damage was to the north and west of us. The carpenters that were doing the remodeling just came the next day and continued their work except with the addition of storm damage. The job was done in a couple of weeks but most people had to wait much longer. There were a couple of strange stories that we encountered with the tornadoes. There was a small Caterpillar bulldozer in front of our house, the gas company was in the process of putting in new lines. The tornado moved the bulldozer 30 feet north of where it had originally been parked. The carpenters had just finished our brick facing in the front of our house, that very day. The mortar was still fresh and moist. The tornado didn't move one brick and that facing is still on our house to this day!

Tom & Gloria Myhra,
6360 Able Street, Fridley

At the time of the tornadoes, I was an English teacher at Fridley High School. I was at the baseball game (Fridley Commons) between Fridley and St. Paul Park. The air was heavy and the sky green. The game wasn't over but I left and my brother from New Brighton called me to say that a tornado was coming! His siren in New Brighton was going but not in Fridley!

Our neighbor and her 3 kids came to our house and we all ran to the basement. It was about 6:30 p.m. and very windy outside. Then, we heard a loud, hard wind sound coming! You could hear lots of debris hitting the siding. We had three tornadoes that night and we took shelter each time. I ended up sleeping on the couch. I grew up in North Dakota where the wind blows all the time and tornadoes didn't really ever materialize but the wind in the Fridley tornadoes was something I will never forget. After the storms, it was really difficult to mow the yard. I was always finding "stuff" when I mowed after that. The gym floor at Fridley High School was raised 6 feet due to the water saturation. We ended up having the graduation at Brooklyn Center High School that year for the seniors.

St. Patrick's Day Blizzard Aftermath, March 17th, 1965

Northern Kan

Unprecedented Spring Flooding, April 20, 1965

Snake River, south of Mora, Minnesota

Record Flooding in St. Paul, April 1965

Shepard Road under water in foreground, High Bridge in the distance

Damage at the 100 Twin Drive-In
at the intersection of Hwys. 65 & 100.

Photo taken by John Croft.
Courtesy of Minnesota Historical Society

"Tornado Over Minnetonka" by H.B. Milligan of Minnetonka. The center of the funnel is near the junction of Hwys. 7 & 101. Winner of the "Best Snapshot of the Week", Milligan won the $5 prize. Courtesy of Minnesota Historical Society

Famous Photographer John Croft took this fantastic picture of the Fridley tornado facing west from Hwy. 65 by Moore Lake. The tornado is completely illuminated by lightning! Courtesy of Minnesota Historical Society

Unbelievable damage from the tornado in Navarre.

Courtesy of Minnesota Historical Society

Cars tossed like toys in the debris. Island Park/ Casco Point-Lake Minnetonka area. Courtesy of Minnesota Historical Society

House completely swept off foundation in Cottagewood. No part of the house was found. Courtesy of Minnesota Historical Society

Unbelievable damage in Mounds View at Lois Drive as the last tornado of the night devastated the area. Courtesy of Minnesota Historical Society

Weather Bureau Teletype Weather Reports- May 6th, 1965

TWIN CITIES WEATHER BUREAU RADAR BULLETIN

MAY 6 9.30AM

STRONG THUNDERSTORM CELLS MOVING TO THE NORTHEAST ARE LOCATED IN THE VICINITY OF ST CLOUD. PRECAUTIONS FOR STRONG WINDS HEAVY RAIN AND POSSIBLE HAIL ARE ADVISED FOR SHERBURNE AND BENTON COUNTIES DURING THE NEXT HOUR..NOT TO BE USED AFTER 10.15 AM

FOLLOWING RELEASED TO PRESS RADIO AND TV AT TWIN CITIES AIRPORT STATION

TWIN CITIES AIRPORT STATION

SPECIAL RADAR BULLETIN ISSUED AT 1030AM CST

AT 1030AM CST WEATHER BUREAU RADAR AT TWINCITIES AIRPORT SHOWS AN AREA OF STRONG THUNDERSTORMS WITH HEAVY RAIN STRONG WINDS AND POSSIBLE HAIL LOCATED FROM 20 TO 40 MILES NORTH OF THE TWIN CITIES MOVING EASTWARD. THE FOLLOWING COUNTIES ARE ADVISED TO TAKE PRECAUTIONS NORTHERN ANOKA ISANTI KANABEC SOUTHERN PINE AND CHISAGO.

THIS IS THE SAME THUNDERSTORM THAT WAS MENTIONED IN THE EARLIER RADAR BULLETIN BUT IT DIRECTION OF MOVEMENT HAS NOW CHANGED FROM NORTHEAST TO EAST.

END FORECASTS 10.40 AM

TWIN CITIES AIRPORT STATION

RADAR BULLETIN 1145 CST

WEATHER BUREAU RADAR SHOWS THE LINE OF THUNDERSTORMS ALONG A LINE FROM ABOUT STILLWATER TO HINCKLEY MUST BEFORE NOON CONTINUING TO MOVE EASTWARD. NO FURTHER STORMS ARE ANTICIPATED IN MINNESOTA BUT FOLLOWING AREAS IN WISCONSIN SHOULD TAKE PRECAUTIONS UNTIL 2PM DAYLIGHT TIME..WASHBURN NORTHERN POLK BARRON AND EASTERN BURNETT COUNTIES. THESE THUNDERSTORMS ARE DECREASING IN INTENSITY SO NO FURTHER BULLETINS WILL BE ISSUED UNLESS FURTHER INTENSIFICATION TAKES PLACE

END

12 NOON TWIN CITIES TEMP 73

RELATIVE HUMIDITY 71 PER CENT

WIND SOUTH 18 WITH GUSTS TO 25 MPH

BAROMETER 29.689 RISING SLOWLY

SKY CLOUDY

END 11.58 AM

NATIONAL WEATHER SUMMARY

MAY 6 152PM

THUNDERSTORMS IN THE MIDWEST AND UNUSUAL COLD IN THE WEST HEADLINE TODAYS WEATHER NEWS.

THE THUNDERSTORMS ARE OCCURRING FROM MINNESOTA TO OHIO AND SOUTHWARD INTO ARKANSAS. ACTIVITY HAS BEEN HEAVIEST SO FAR TODAY IN CENTRAL MINNESOTA NORTHEAST WISCONSIN AND NORTHERN MICHIGAN. ALPENA IN NORTHEAST LOWER MICHIGAN WAS SWAMPED BY ALMOST TWO INCHES OF RAIN IN THE SIX HOUR PERIOD ENDING AT NOON.

PARTS OF MINNESOTA AND NORTHWEST WISCONSIN WERE UNDER ALERTS FOR HEAVY THUNDERSTORMS AND HAIL AT VARIOUS TIMES TODAY. THE STORMS FOLLOWED A NIGHT OF TORNADOES IN THE UPPER MIDWEST. AROUND A DOZEN TWISTERS SKIPPED ACROSS RURAL AREAS OF MINNESOTA WISCONSIN IOWA AND THE DAKOTAS LATE WEDNESDAY AND WEDNESDAY NIGHT.

IN THE WEST A WARNING TO STOCKMEN IS IN EFFECT FOR THAT PORTION OF MONTANA LYING EAST OF THE CONTINENTAL DIVIDE. COLD RAIN OR SNOW IS EXPECTED OVER EASTERN AND SOUTHERN MONTANA TODAY AND TONIGHT WITH LOCALLY HEAVY SNOW ABOVE THE 5000 LEVEL NEAR THE WYOMING BORDER. DILLON IN SOUTHWESTERN MONTANA MEASURE 7 INCHES OF SNOW ON THE GROUND AT NOON WHILE LEWISTOWN IN THE CENTRAL PART OF THE STATE REPORTED 2 INCHES. A FREEZE WARNING HAS BEEN ISSUED FOR UTAH AS TEMPERATURES OVER MUCH OF THE WEST REMAIN BELOW NORMAL.

IN CONTRAST TO THE CHILLY CONDITIONS IN THE WEST WARM AIR IS STREAMING NORTHWARD FROM THE GULF OF MEXICO ACROSS THE BULK OF THE CENTRAL AND EASTERN STATES.

NATIONAL WEATHER SUMMARY

U.S. WEATHER BUREAU

THURSDAY MAY 6 1965

TIME ... 752PM CST

SEVERE WEATHER FROECAST AREAS DOT

THE NATIONS MIDSECTION THIS EVENING.

WARM AND HUMID WEATHER CONTINUES TO SPREAD NORTH ACROSS THE CENTRAL PLAINS AND THE UPPER MISSISSIPPI VALLEY. SEVERE THUNDERSTORMS DEVELOPED IN OKLAHOMA AND IN THE UPPER MISSISSIPPI VALLEY EARLY THIS EVENING AND AS LAST NIGHT ARE SPAWNING TORNADOES WITH LITTLE OR NO EFFORT.

TORNADOES HAVE BEEN REPORTED 35 TO 40 MILES WEST OF MINNEAPOLIS MINNESOTA AND SEVERAL TORNADOES HAVE BEEN SIGHTED IN EXTREME WESTERN OKLAHOMA.

THE WARM AND HUMID WEATHER IN THE SOUTHEAST HALF OF THE COUNTRY IS CONTRASTED BY THE COLD WEATHER IN THE NORTHWEST HALF

OF THE COUNTRY. FOR EXAMPLE LEWISTOWN IN CENTRAL MONTANA REPORTED FOUR INCHES OF NEW SNOW THIS AFTERNOON FOR A TOTAL OF SIX INCHES ON THE GROUND AT 6PM CST TODAY.

PRECIPITATION HAS BEEN LOCALLY HEAVY IN THE GREAT LAKES REGION. OF THE HEAVIER AMOUNTS SAGINAW IN EAST CENTRAL LOWER MICHIGAN REPORTED NEARLY ONE INCH/.989/SINCE MID DAY.

MAJOR MAP FEATURES AT 6PM CST THURSDAY...A WARM FRONT EXTENDS FROM NOTHER CAROLINA TO NORTHEAST MICHIGAN. A COLD FRONT EXTENDS FROM UPPER MICHIGAN TO A LOW/29.56/ IN NORTHEAST NEBRASKA WITH THE COLD FRONT TO SOUTHWEST TEXAS. ANOTHER LOW/29.41/ IS CENTERED IN EASTERN COLORADO WITH A COLD FRONT TO SOUTH CENTRAL ARIZONA.

WEATHER STORY PREPARED AT THE TWIN CITIES WEATHER BUREAU A COLD FRONT IS OVER MINNESOTA BEGINNING NEAR LAKE SUPERIOR AREA AND EXTENDING SOUTHWESTWARD THROUGH THE TWIN CITIES AND INTO NORTHEAST IOWA. HEAVY THUNDERSTORMS DEVELOPED ALONG THIS LINE IN

MINNESOTA THIS AFTERNOON AND EVENING.

NUMEROUS FUNNEL CLOUDS AND TORNADOES WEST AND SOUTHWEST OF MINNEAPOLIS DEVELOPED AND MOVED EAST NORTHEAST OVER THE NORTHERN HALF OF MINNEAPOLIS ST. PAUL. TWO CONFIRMED TORNADOES OCCURRED ONE EAST OF EXCELSIOR NEAR THE INTERSECTION OF HIGHWAYS 101 AND HIGHWAY 7 AT 7 PM. THE SECOND TORNADO WAS IN THE FRIDLEY SPRING LAKE PARK AREA AT 7.15PM THE NUMBER OF INJURIES OR DEATHS FROM THESE STORMS IS NOT KNOWN.

END 9.02PM

9 PM TWIN CITIES TEMP 70

WIND SOUTHWEST 14

HUMIDITY 87

BAROMETER 29.16 RISING

WEATHER THUNDERSTORM WITH
LIGHT RAIN SHOWERS

END 9.25PM

The Infamous 1965 Spring Floods of Minnesota

People have long talked about the possible connection of the record cold winter and spring flooding which may have lead to the record tornado outbreak of May 6th, 1965. Bitter cold in the month of November 1964 followed by rain storms which turned to ice had made the ground rock hard. This foundation made the flood run-off even more record breaking when the second coldest, snowiest March in history came in 1965. A record 36" snowfall was recorded by St. Cloud on the St. Patrick's Day blizzard. This blizzard shut down schools for the first time since 1950. Winds were clocked at 30-75 miles per hour and the temperatures plunged below zero in the days after. Daily temperatures averaged 17.1 degrees-8 degrees below normal setting yet another all-time record. As early as March 25, St. Louis Park engineer Ray Folland expressed concern about flooding if all that snow melted too fast. This prediction unfortunately came true in April 1965.

Heavy rainfalls in early to mid April, falling upon the snowpack and frozen grounds, with nowhere to go due to the frozen ground, the rain and melted snow quickly found their way into the Mississippi River and its tributaries. This runoff created the worst flooding in Minnesota history.

In the Mankato area, the melting snow drained down to the Minnesota River, raising it. Snow melted from other towns upstream from Mankato flowed downstream into Mankato's section of the river. Other rivers such as Indian Creek and the Blue Earth River had a fast thaw and their water flooded over into the Minnesota River. When the flooding hit Mankato, people tried hard to stop the water from destroying the whole city. Volunteers came from all over to help sandbag the rising waters. Over 2000 people came to lay down a million sandbags. Called the "Great Wall of China", the wall was over 3000 feet long and 6 feet high and took only 36 hours to construct. Yet that still wasn't enough. Evacuation of the lower areas of Mankato and West Mankato was ordered but North Mankato escaped the floods because they had already

built a large wall on their side of the river.

A 672-mile stretch of the Mississippi River (from 100 miles north of Minneapolis to Hannibal, Mo.) experienced the worst flooding in history, lasting the entire month of April. In some places, the flooding went on for up to 43 days and 16 persons lost their lives to this historic flood. U.S. Weather Bureau Gage Height records began in 1873 and this flood is still considered a "once in a century occurrence".

Here is some interesting data from the LaCrosse Historic Gage records of the top five floods since 1873.

#1: April 20, 1965
Crest 17.9 feet
(velocities in the channel 6 feet/second)

#2: June 19, 1880
Crest 16.5 feet
(10 day rain deluge)

#3: April 20, 1969
Crest 15.7 feet

#4: April 20, 1952
Crest 15.3 feet

#5: April 19, 1951
Crest 14.9 feet

In the Twin Cities, persons prepared for the onslaught of the convergence of three swollen river systems: Minnesota, Mississippi, and St. Croix. Snowfall in March was over 300% above normal in east and east-central Minnesota.

Timeline of the flood in Twin Cities.

April 2, 1965:

35/100 inch of rain following 10 inches of snow that fell on March 31. High temperature of 42 degrees and the flood crest predicted at 12 feet by April 9th or 10th.

April 4, 1965:

Rain and drizzle continue, increasing flood danger with the four river valleys bracing for floods in Minnesota.

April 6 & 7, 1965:

Heavy rains occur in flood affected areas.

April 12, 1965:

39 counties proclaimed disaster areas.

April 13, 1965:

Flood tops records at 22.95 foot crest. The Mississippi is increasing at 1/10 ft. per hour.

April 15, 1965:

President Johnson visits Minnesota flood damage. Crest at 25.35 feet.

April 16, 1965:

Good Friday cold snap delays Mississippi River crest. Crest at 25.9 feet.

April 17, 1965:

Crest at incredible 26 feet, snow expected that night.

The spring Mississippi River flood of 1965 stands as the flood of record for nearly half of the river's length. At the time, the crests of that April exceeded previous records by several feet at many river gage sites. To this day, those record crests still out distance the 2nd highest crest by a foot or more at many of those same sites. The flood caused $225 million in damage to public and private properties, with $173 million of that occurring along the main stem of the Mississippi River. Emergency actions and evacuations, based on Weather Bureau forecasts and excellent reporting, prevented approximately $300 million in additional damages.

Excerpts from Weather Bureau article in St. Paul Dispatch: May 12, 1965 by R.J.R. Johnson

Jim Stone, Weather Bureau meteorologist, had been watching an ominous thunderstorm grow and intensify on his radar screen since around 4:30 p.m. on May 6th. The storm moved in from southwest of the Twin Cities, steadily increasing in size and intensity. By 5:43 p.m. it was in the southwestern suburbs. The Kansas City Weather Bureau at 5:45 p.m. issued a severe weather forecast for southern Minnesota. It warned of the possibility of tornadoes. At 6:00 p.m. the Weather Bureau here issued a bulletin: radar showed heavy rain, thunderstorms and hail from the Twin Cities to 100 miles westward. The storm was moving northeastward. "It didn't look too threatening until just before 6," said Niel Lampi, forecaster in charge at the Twin Cities Weather Bureau station that evening. At 6:20 p.m. Excelsior reported hail an inch in diameter. Just about 6:30 p.m. Stone saw what he'd been waiting for and afraid of finding. A sharp hook image appeared jutting from the storm shadow on his screen. It was in the vicinity of Chanhassen. To Stone and Lampi, the storm was an old acquaintance. They'd been tracking it for two hours. The northeasterly path would carry it right across the northwestern edge of the Twin Cities metropolitan area. They agreed. This was the time to initiate the take-cover civil defense sirens. "I didn't really hesitate a bit," Lampi said later of his decision. It was the first time the air raid alarm was used as a tornado warning since the system was instituted in May, 1959. There was a moment lost when the lock on the alarm dial resisted. Then Lampi reached for the take-cover slot on the telephone-like dial and turned it twice. It was an act that later was credited with saving hundreds and perhaps thousands of lives. All over the metropolitan area sirens began to wail. Fear became widespread as people realized it was not a test, or a short-circuit in one siren. A tornado warning tape was fed into the teletype machine at the Weather Bureau and tapped out its message in newsrooms throughout the area. Lampi hit the bell key over and over. Remembering constant civil defense advice, people turned mostly to their radios to find out what was happening. The news was grim and it was to grow grimmer throughout the evening. Weather Bureau radar sightings and citizen reports were broadcast. The Weather Bureau kept up a steady stream of teletype chatter, reporting progress of the storm. What had started out as one tornado, a very destructive one, grew to a total of seven in the immediate area with two others outstate.

Here is a breakdown of tornado occurrences including the controversy of the three tornadoes to strike Fridley that evening.

- **Tornado No. 1** hit Chanhassen at 6:27 p.m., was tracked as it crossed hwys. 101 & 7 and passed through Golden Valley and Robbinsdale. It moved at a nearly steady 38 miles per hour and smacked into the Northern Ordinance plant in Fridley. The Fridley city hall clock stopped at 7:10 p.m.
- **Tornado No. 2 and 3** tore into Fridley; at 8:10 p.m. and 9:15 p.m. A powerful straight wind hit a trailer park there at 10 p.m.
- **Tornado No. 4** hit Green Isle at 6:50 p.m. moving on to Hamburg, Norwood, and Island Park. The clock of a Green Isle farmer who died stopped at 6:55 p.m.

About the same time,

- **Tornado No. 5** was skipping from Glencoe to Lester Prairie.
- **Tornado No. 6** moved eastward from Hamburg to Gotha.
- **Tornado No. 7** appeared in the Lake Elmo-Hugo area at 9:15 p.m.
- **Tornado No. 8** was at Clear Lake, southeast of St. Cloud at 8:03 p.m.
- **Tornado No. 9** was at Mont-gomery at 8:17 p.m.

At the Bureau everyone was jumping. John Bottomley, Joe McBride and John Parry, aviation weather forecasters, pitched in to help with telephone calls; Dean Nesley, weather observer, became a general hand. Bud Jenkin-son, radar technician, fought occasional power lags to keep the radar going, and Vinton Bous-lough, public service man, kept the press, radio and television informed. P. W. Kenworthy, me-teorologist in charge of the Twin Cities station, and his chief assistant, Joseph Strub Jr., arrived from home to help out. "Once we started go-ing," Lampi said, "I had a feeling we were dealing with something big. The next night, Friday, we spotted some hooks south of here, but the situation wasn't the same. Friday didn't scare me the way Thursday did." No one will ever know how many more lives would have been lost if Lampi had not activated the sirens. Perhaps those people who came alive out of basements have an idea. "We think these men did an excellent job." Kenworthy said. "They kept calm and acted like real veterans." Shrugging off Kenworthy's praise, Lampi said, "We just happened to be the ones on duty."

Hennepin County Sheriff Radio Transmission Records

The following is a resume of the events pertaining to the evening of May 6, 1965 Tornadoes that occurred in the Metropolitan and Lake Areas on the Hennepin County Sheriff Radio Systems. It should be noted that the chronological times are in P.M. and these are the actual transmissions of the various squads in the area. The "#" notation refers to the highway number.

6:19	Car 840	Reported Hail in Excelsior.
6:30	Car 840	Hail as big as Golf Balls.
6:32	Car 811	Pole on Fire #7 & St. Alban's Bay.
6:34	Car 710	Tornado over Minnetonka & #7
6:34	Tornado looks like over St. Alban's Bay. Heading from Excelsior on #82.	
	Car 810	Went thru Deephaven lots of damage, call Northern States. Send Help.
	Car 811	Trees all down Navarre & Shorewood.
	Car 710	Will get in to help.
	Car 733	Heading for Area.
6:43	Alerted all Fire Stations & Personnel.	
6:53	Smith Ambulance ordered to scene.	

7:03 Delano 820 Tornado hit Glencoe-
heading this way.

7:09 Sigalert-for CD volunteers.

7:19 Fridley Car 931-City Hall hit Tornado-
Need lots of help.

7:38 Car 840 Another Funnel sighted here.

7:41 Car 835 Max Pizza, now coming this way.

7:42 Car 835 Coming thru Wayzata.

7:42 Car 811 Set up Minnetonka High.

7:43 Car 843 Loring Acres-Houses down.
Island Park hit-need help.

7:50 P.M. Tornado at 6 & 101 heading north-toward
Corcoran area 835. Tornado sighted at 86th Penn
So. Another at #55 & Fernbrook.

8:14 P.M. Tornado North of Hopkins, heading North.

Car 931 Car hit tree
40th & Marshall St.-Tornado
hit Winnetka & #55-
Another one 101 & East of #52-
Okay-

833 12 & 100 hit
Bad- Ewald Terrace house
over.

8:32 P.M. County 18 between 77th and 82 St.-
Power poles down all over.

8:54 P.M. 931-60th University-Second twister going

through-30 People trapped
At Midland-Coop-power
out at Fridley station-Fire station hit-no police
radio-send car for communication-
car 778 will go to police station.

Foregoing was some of the activity recorded at the Radio Station. Approximately five hours of solid recordings were made at the Radio Station of the tornado operation initially.

When the first tornado alert was received by two-way radio from Excelsior, car 840, all units of the Sheriff's system, the Metro-alert system and the Fire Channel were alerted by the radio dispatchers. At the same time, the dispatchers were alerted by the "Take cover alert" on the Bell and Light systems. When this happened, the telephone systems were jammed with incoming calls of inquiry about the Air Raid Siren soundings.

There were two dispatchers and one technician on duty. An emergency squad member arrived at radio to assist in the operation. Several more squad members arrived shortly. A half hour later the Radio Supervisor, Joe Sentyrz, was at radio. Mr. Halstead, CD director, obtained a walkie-talkie unit at radio and proceeded to the Minnetonka High. Mr. Newstrom, of Budget and

Purchasing, arrived at the height of the first tornado alert and assisted in dispatching the General Hospital Ambulances on the hospital radio system. North Memorial and Smith ambulances were dispatched. Captain Mair was at radio for a short time and then left for the disaster area. All personnel at radio remained at their posts during the twister movements even though the reports said that it was heading for the station.

Sgt. Seviola stopped to report that he was blown off the road by the twister when the first tornado alert was received by two-way radio from the Excelsior, car 840, all units of the Sheriff's system, the Metro-alert system and the Fire Channel were alerted by the radio dispatchers. At the same time the dispatchers were alerted by the "Take cover alert" on the Bell and Light systems. When this happened the telephone systems were jammed with incoming calls of inquirer about the Air Raid Sirens soundings.

Captain Mair was at radio for a short time and then left for the disaster area. All personal at radio remained at their posts during the twister movements even though the reports said that it was heading for the station. Sgt. Seviola stopped to report that he was blown off the road by the twister on #55. About four hours later Chief

Smith and some of his Emergency personnel stopped at radio and relieved the telephone and radio operators. They then left for the Fridley area. Telephones were answered at a rate of one every two seconds throughout the evening and night. Emergency calls were honored, inquiries were rerouted. When on phone line was cleared, another call was coming in.

The Fridley Police communications were out because of power outage and car 778, Deputy Ringate, was sent to the Police Station to establish communications with us. Both Fridley Police cars with radios on our systems were hit by the storm and were incapacitated.

Deputy Bill Berry and Bill Gillespie set up a command post at Navarre and directed operations on Channel Number 2 in the area. The Boat Patrol equipment was installed at the Deephaven Hall and operated by Captain Williams and his men for Shorewood, Excelsior, Deephaven, and surrounding areas.

The Mobile Amatcur Radio Corps (MARC) was alerted when the first call came of a possible Tornado in this area from the WCCO monitor. They were sent to Navarre at the direction of Mr. Halstead to set up communications there. Later, the communications truck was transferred to Spring Lake Park by Mr. Halstead. Saturday night Chief McCarthy of Fridley, requested their services to cruise the Fridley area during the dark hours to spot and report fires and other emergencies back to police headquarters. Two nights of operations were necessary until temporary lines were installed.

Portable generating plants were located and sent to all areas without power. A number of Fire Departments with generating plants stood by at the scenes for extended times. After a catastrophe, it is imperative to evaluate what was done and make recommendations for future operations so that preparations can be made to handle the situation more speedily and efficiently.

Village of Orono: Report on Activity at Tornado Disaster Area-May 13, 1965 by Chief of Police, Jerry Ross

This tornado hit the Navarre and Casco Point areas at approximately 7:36 p.m. on May 6, 1965. We arrived on the corner with both cars at about 8:00 p.m., same date. Officer Krotzer was in the one car and I was in the other, along with two of our Civil Defense personnel. By this time, ambulances had been ordered for the area and were on the way. (From here on, for the rest of this evening, I was too busy to keep track of the time.)

The Navarre area was a shambles and when we arrived most of the injured had been already transferred to the Navarre Inn. We checked these injured first and as soon as the ambulances arrived we sent what injured were there to the hospital. At the same time we were also establishing security in the area. Then Officer Krotzer started in through the residential area, with some ambulances along with him, and they took out what injured they found down there. By this time the Maple Plain and Wayzata Fire

Departments were there and were helping with the injured also. As soon as we thought we had all the injured out, the Maple Plain Fire Department used chain saws and started clearing the roads. By this time Officers Dressol and Burmaster were also on the scene along with some County personnel.

I then sent our officers to make another check in the residential area to see if we missed any injured. They checked with all the neighbors and as near as they could tell, everybody was accounted for. I believe we had approximately 15 injured altogether. During this time we got some portable light plants on the corner to try to keep the area under surveillance as best we could. There were contractors there, bringing in plywood to nail up where the windows wore out of the buildings, and this operation went on until they finished, which was just about morning.

As soon as it got light I sent our officers to make a further check on the residential area so we could be positive that we had not missed any casualties. We once again accounted for everybody that was supposed to be in the area, and apparently we got them all, in the night, to the hospitals.

At this time we started thinking about security for the weekend and for whatever length of time that it was going to be needed, and it was about this time also that Sergeants Berry and Gillespie from the Hennepin County Crime Lab arrived to help me. I believe their own report that will be accompanying this report will explain how the operation was handled from this point on.

I would like to express my thanks to the Council and to the Administrator for allowing me to handle the disaster situation without interference. I know we made some mistakes in handling this situation but, it is something one does not get experience in until it happens. I wish to be sure that Council understands how much assistance I received from Sergeants Berry and Gillespie, and also the tremendous amount of time put in by our own personnel.

Original Fujita Tornado Scale

The Fujita scale (F-scale), or Fujita-Pearson scale, is a scale for rating tornado intensity, based on the damage tornadoes inflict on human-built structures and vegetation. The official Fujita scale category is determined by meteorologists and engineers after a ground and/or aerial damage survey; and depending on the circumstances, ground swirl patterns, radar tracking, eyewitness testimonies, media reports and damage imagery.

The scale was introduced in 1971 by Tetsuya "Ted" Fujita of the University of Chicago who developed the scale together with Allen Pearson (path length and width additions in 1973), head of the National Severe Storms Forecast Center in Kansas City, Missouri. The scale was applied retroactively to tornado reports from 1950 onward for the National Oceanic and Atmospheric Administration Tornado Database in the United States, and occasionally to earlier

infamous tornadoes. The Twin Cities Tornado Outbreak of May 6th, 1965 was one of those weather events.

▽ **F0: 40-72 mph**

Minimal Damage:

Some damage to chimneys, TV antennas, roof shingles, tree branches, and windows.
Frequency: 29%.

▽ **F1: 73-112 mph**

Moderate Damage:

Automobiles overturned, carports destroyed, trees uprooted.
Frequency: 40%.

▽ **F2: 113-157 mph**

Major Damage:

Roofs blown off homes, sheds and outbuildings demolished, mobile homes overturned.
Frequency: 24%.

▽ **F3: 158-206 mph**

Severe Damage:

Exterior walls and roofs blown off homes. Metal buildings collapsed or are severely damaged. Forests and farmland flattened.
Frequency: 6%.

▽ **F4: 207-260 mph**

Devastating Damage:

Few walls, if any, standing in well-built homes. Large steel and concrete missiles thrown far distances.
Frequency: 2%.

▽ **F5: 261-318 mph**

Incredible Damage:

Homes leveled with all debris removed. Schools, motels, and other large structures have considerable damage with exterior walls and roof gone. Top stories demolished.
Frequency: Less than 1%.

The severity and sheer number of major, severe, and devastating tornadoes the evening of May 6th, 1965 is monumental and statistically, extremely rare. In the swarm of tornadoes that hit the Twin Cities that evening, there was (1) F2, (1) F3, and (4) F4 level tornadoes.

The Enhanced Fujita Tornado Scale

The Enhanced Fujita Scale, or EF Scale, is the scale for rating the strength of tornadoes in the United States estimated via the damage they cause. Implemented in place of the original Fujita scale introduced in 1971, it began operational use on February 1, 2007.

The scale has the same basic design as the original Fujita scale, six categories from zero to five representing increasing degrees of damage. It was revised to reflect better examinations of tornado damage surveys, so as to align wind speeds more closely with associated storm damage. Better standardizing and analyzing what was previously subjective and ambiguous, it also adds more types of structures as well as vegetation, expands degrees of damage, and better accounts for variables such as differences in construction quality.

As with the Fujita scale, the Enhanced Fujita Scale remains a damage scale and only a proxy for actual wind speeds. While the wind speeds associated with the damage listed have not undergone empirical analysis owing to excessive cost, the wind speeds were attained through a process of expert elicitation based on various engineering studies since the 1970's as well as from field experience of meteorologists and engineers.

The new scale takes into account quality of construction and standardizes different kinds of structures. The wind speeds on the original scale were deemed by meteorologists and engineers as being too high and engineering studies indicated that slower winds than initially estimated cause the respective degrees of damage. Essentially, there is no functional difference in how tornadoes are rated. The old ratings and new ratings are smoothly connected with a linear formula. The only differences are adjusted wind speeds, measurements of which weren't used in previous ratings, and refined damage descriptors; to standardize ratings and to make it easier to rate tornadoes which strike few structures. The new scale lists an EF5 as a tornado with winds at or above 200 mph, found to be sufficient to cause that damage previously ascribed to the F5 range

of wind speeds. In other words, many of the F4 tornadoes of May 6th, 1965 would now be rated as EF5's due to the unbelievable damage caused by these storms. The six main categories for the EF Scale are listed below, in order of increasing intensity. The damage of the Twin Cities tornadoes can be transferred from the original Fujita scale to the EF scale due to associated storm damage. It should be noted that tornadoes recorded on or before January 31, 2007 will not be re-categorized as was done with the Fujita scale in 1971.

The six categories for the EF scale are listed below, in order of increasing intensity.

- **EF0: 65-85 mph/105-137 km/h**

  Light Damage:

  Peels surface off some roofs; some damage to gutters or siding; branches broken of trees; shallow-rooted trees pushed over.

- **EF1: 86-110 mph/138-178 km/h**

  Moderate Damage:

  Roofs severely stripped; mobile homes overturned or badly damaged; loss of exterior doors; windows and other glass broken.

- **EF2: 111-135 mph/179-218 km/h**

  Considerable Damage:

  Roofs torn off well-constructed houses; foundations of frame homes shifted; mobile homes completely destroyed; large trees snapped or uprooted; light-object missiles generated; cars lifted off ground.

- **EF3: 136-165 mph/219-266 km/h**

  Severe Damage:

  Entire stories of well-constructed houses destroyed; severe damage to large buildings such as shopping malls; trains overturned; trees debarked; heavy cars lifted off the ground and thrown; structures with weak foundations blown away some distance.

- **EF4: 166-200 mph/267-322 km/h**

  Devastating Damage:

  Well-constructed houses and whole frame houses completely leveled; cars thrown and small missiles generated.

- **EF5: >200 mph/>322 km/h**

  Incredible Damage:

  Strong frame houses leveled off foundations and swept away; automobile-sized missiles fly through the air in excess of 109 yd (100 m); high-rise buildings have significant structural deformation; incredible phenomena will occur.

Three Tornado Controversy

It has been established that the evening of May 6th, 1965 produced at least 24 "hook-shaped" echoes on Weather Bureau radar systems. Many instances of clocks stopping when a tornado hit that specific location have been documented even though the Weather Bureau doesn't acknowledge the existence of that tornado in that area at that specific time frame. Many eyewitness reports of tornadoes extending to ground were reporting live on WCCO radio that evening giving more credence to the theory of many more tornadoes than the reported six confirmed tornadoes previously described. These details along with reliable witnesses give the three-tornado event of Fridley a very credible likelihood of occurrence.

State climatologist Earl Kuehnast pro-

duced a map of tornado tracks in the 1960's that showed three tornadoes passing through Fridley on May 6th. For eight years the records remained the same. But in 1973, he produced new maps listing different tornado tracks and times. The new maps showed only two tornadoes passing through Fridley, the first one began at 7:06 p.m. and the other at 8:14 p.m. Todd Krause, a Fridley resident who works as a warning coordinator meteorologist for the National Weather Service, said the tornado that touched down at 6:27 p.m. in Chanhassen was originally thought to have traveled through Fridley, but the updated records show that tornado ended at 6:43 p.m. in Deephaven.

Kuehnast's new maps were approved by Weather Bureau Principal Assistant Joseph Strub Jr. who later became meteorologist-in-charge in 1969. Two maps published in 1975 show the revised information and it was put in the national database without documentation or justification for the change. The archived records simply show X's made by Kuehnast on the 1965 version of tornado tracks showing three tornadoes passing through Fridley.

Forty-two years later, the most intriguing question concerns the 1973 correction to the original data assigned in 1965. Tornado tracks and times was changed, and the number of fatalities was reduced from 14 to 13. Today, there is no information at the National Weather Service or State Climatology as to why the data were revised. Three hand-sketched maps showing the new data were uncovered from the files of Earl Kuehnast and they were dated January and February 1973. Kuehnast even placed a large "X" across the 1965 version clearly indicating he no longer believed the original map to be correct. Two maps published in 1975 show the revised information-these are in "Climate of Minnesota, Part VIII-Precipitation Patterns in the Minneapolis-St. Paul Metropolitan Area and Surrounding Counties." And can be seen at http://climate.umn.edu/pdf/climate_of_minnesota/comVIII.pdf, pages 30-32.

The updated records now show six tornadoes passing through the Twin Cities the night of May 6th, 1965. The old Fujita scale (1971-2007) was used in calculations of damage caused by the listed tornadoes.

The following are now the official NOAA records:

- **Tornado #1** touched down at 6:08 p.m. (CST) just east of Cologne in Carver County, was on the ground for 16.5 miles traveling at approximately 31 miles per hour. It dissipated in the northwestern portion of Minnetrista in Hennepin County at 6:40 p.m. This tornado was rated an F4, killed 3 people and injured 175.
- **Tornado #2** touched down at 6:27 p.m. (CST) near Lake Susan in Chanhassen in Carver County and traveled 8 miles straight north at approximately 30 miles per hour. It dissipated in Deephaven in Hennepin County at 6:43 p.m. This tornado was rated an F4, but fortunately no injuries or fatalities reported.
- **Tornado #3** touched down at 6:34 p.m. (CST) about 3 miles east of New Auburn in Sibley County traveling 15 miles northeast at approximately 33 miles per hour. It dissipated just west of Lester Prairie at 7:01 p.m. This tornado was rated an F3 with no reported injuries or fatalities.

- **Tornado #4** touched down at 6:43 p.m. (CST) about two miles east of Green Isle in Sibley County and was on the ground 12 miles at approximately 31 miles per hour. It dissipated about 2 miles southwest of Waconia in Carver County at 7:04 p.m. This tornado was rated an F2, killed one person and injured 175 people.

- **Tornado #5** touched down at 7:06 p.m. (CST) in the southwestern most corner of Fridley in Anoka County, moved across the Northern Ordinance plant traveling 6 miles at approximately 24 miles per hour. It dissipated just northeast of Laddie Lake in Blaine at 7:21 p.m. This tornado was rated an F4, killed 3 people and injured 175.

- **Tornado #6** touched down at 8:14 p.m. (CST) in Golden Valley in Hennepin County, moved across north Minneapolis, Fridley, and Mounds View for 20 miles at approximately 25 miles per hour. It dissipated just west of Centerville at 9:02 p.m. This tornado was rated an F4, killed 6 people and injured 158.

Original 1965 Weather Bureau Report

A family of 6 tornadoes struck an 11 county region with greatest damage and concentration southwest and just north of Minneapolis. Through the evening, 24 hook shaped echoes appeared on the Minneapolis WBAS radar and a large number of confirmed and unconfirmed funnels reported. Many reports of hail, up to golf ball size, throughout the region.

The first touchdown occurred at a farm 1 mile south of Chanhassen. The tornado then skipped northward to Chanhassen hitting a lumberyard and 6 buildings in a shopping center in a path 150 yards wide. Two clocks in the vicinity stopped at 6:27 p.m. From here northward to Lotus Lake, about 20 homes received major damage or destroyed with 15 homes having minor damage. The tornado continued northward into the Christmas Lake where 18 homes destroyed and 32 received major damage. It then moved over Deephaven damaging 100 homes, curved northeastward, thereby going south of Medicine Lake, touched down briefly at Hwy. 100 in Golden Valley where it inflicted major damage on at least six homes. Several homes in Minneapolis were damaged before it moved over the Mississippi River into the lower section of the suburb of Fridley. Here it traveled northward hitting the Northern Ordinance Division of the FMC Corporation at 7:05 p.m. continuing through Fridley. At Fridley City Hall the clock stopped at 7:10 p.m. giving the time of the first of three tornadoes to hit this suburb. Many Fridley residents hit by 2 tornadoes in an hour and 35 minute period; some suffered damage from all three. Excellent warnings by the Weather Bureau and fast dissemination by radio held the Fridley death count to two people. Damage from the three tornadoes in Fridley included 425 demolished and 1100 badly damaged homes. About 1700 people had to leave their homes and find temporary shelter elsewhere. The Fridley school district superintendent estimated the damage to schools alone at about $5 million. Private prop-

erty damage estimated by the mayor at $8.5 million. Twenty-thousand phones were out of service during the night in the northern suburbs. After Fridley, this first tornado skipped northward passing just west of Spring Lake Park and touching down near Blaine, severely damaging Anoka County Airport and destroying 10 planes.

A second tornado reported at 6:50 p.m. near Green Isle by an airline pilot. It moved northeastward flattening farm buildings and trees. Three miles SE of Hamburg a farmer killed along with a barn full of prize cattle at 6:55 p.m. The funnel cut a ? mile wide swath through here. It continued northeastward just missing Norwood-Young America, and Waconia in a path of snapped, twisted, and stripped trees that sometimes was as wide as ? to ? of a mile. All buildings completely destroyed on 20 farms in this 8 mile long section of the track. The vortex moved upward and descended again near Hwy. 7, west of Lake Minnetonka. It moved over Island Park, killing one woman, and destroying 8 homes and 3 farm building units. It continued through Mound killing 2 people and damaging many homes. Moving through Spring Park and Navarre it destroyed about 30 homes, many boats and a resort, and badly damaged the Navarre business district. It crossed Hwy. 12, just west of Wayzata, destroying the Country Club horse barn. From here it moved into the Hamel area where it apparently dissipated.

The third tornado began about 6:55 p.m. midway between New Auburn and Glencoe and moved about 15 miles to the northeast dissipating near Lester Prairie. It followed a zigzag pattern, destroying or badly damaging 25 farm building units and demolishing a church and school. Damage estimate placed at about $1 million.

The fourth tornado touched down about 7:15 p.m. and followed a short path from southwest of Cologne (west of Gotha) to just northeast of Cologne. Three barns were destroyed.

The fifth tornado first observed near the Minneapolis-Fridley border at 37th Avenue NE and Marshall. Several people had observed a huge funnel cloud form in the Chanhassen area sometime before and move northward. This may have been the beginning of this tornado. It moved from northwestern Minneapolis eastward over Northern Ordinance (the second strike on this large industrial complex) at 8:10 p.m. and then NNE across Fridley. It hit Fridley's main school and park complex, among other struc-

tures, then moved northeasterly and struck the Fridley Trailer Court. This and the sixth tornado both hit this 275 unit trailer court. Two-hundred trailers were destroyed, some apparently had exploded, but only one person, an infant killed. Many people in this Park took shelter in the cinderblock wash house. The tornado continued northward from here into the next suburb, Spring Lake Park. Here 150 homes destroyed and 145 heavily damaged. About 75% of Spring Lake Park businesses in the city were demolished and 900 people left homeless. The tornado lifted or dissipated at this point, although sometime later there was wind damage to 3 farms in Braham area, 50 miles NNE of Spring Lake Park.

The sixth tornado first touched down in Golden Valley near 8:40 p.m. at Douglas Drive and Hwy. 55 damaging 25 homes and 8 businesses. It traveled northeastward touching down again in Fridley (the third tornado to hit Fridley) near Hwy. 100 and University, moved northward through Fridley between the paths of tornado number one and number five. It curved ENE at Mississippi Street, then turned northward again just west of Hwy. 65 and went through the Fridley Trailer Court. It moved northeastward through the suburb of Mounds View where 6 people killed and 150 injured. At least 45 homes and 6 apartment buildings destroyed. Total damage in Mounds View amounted to about $1 million dollars. The tornado continued moving northeastward passing over Centerville Late at 8:57 p.m. Four farms north of here damaged in a ? mile wide path. It then continued moving northeastward as a funnel cloud touching down again briefly in the Almelund area at 9:20 p.m. where it snapped trees and damaged buildings on 3 farms.

Weather Reports for May 2rd-May 7th, 1965

May 2, 1965-Sunday:

Cooler air moved into North Dakota and South Dakota, while unseasonable warm temperatures continued in Wisconsin and southern Minnesota. Actual high/low: 85/54 with a trace of precipitation. A surface Low was located in the 4 corners states being held in check for the stronger weather system to move out east.

May 3, 1965-Monday:

A stationary front with a low located north of International Falls is bringing considerable cloudiness and cooler temperatures with showers in central and southern Minnesota. Actual high/low: 69/51 with .08″ precipitation. Note: The 30-day forecast for the region is above average precipitation with below average temperatures. Weak High pressure was moving to the east setting the stage for the stronger Low pressure in the four corners to dominate the local weather.

May 4, 1965-Tuesday:

HEADLINE-"Hail, Wind Plague Region; Scattered Rain Due Today". Hail and thunderstorms pelted parts of southern Minnesota and western Wisconsin with strong winds wrecking several farm buildings near Truman, Minnesota early Monday. Some storekeepers had to clear hail drifts from their doorways. The forecast is for cloudy skies during the day followed by showers or thunderstorms late tonight and Wednesday. A cold front was passing through SE Minnesota. Actual high/low: 62/49 with no precipitation. A huge bubble of warm, moist air was poised just to the south providing a high likelihood of surface based tornadoes in the following days.

May 5, 1965-Wednesday:

A warm front is approaching from the Dakotas and the forecast is for mostly cloudy skies with showers. Cooler temperatures are expected. A low pressure area expanding from the Rockies is expected to bring showers and thunderstorms into the Upper Midwest. Actual high/low: 73/52 with .08" precipitation. The weak High pressure that had kept the warm, moist air to the south shifted straight east thus opening the door for the strong Low to come in the back

of it. The warm front to the south now moves north into Minnesota setting the stage for record tornadoes.

Note:

April Weather Calendar states the actual:

April High/Low: 76/26

April Average Temperature/Normal: 41.8/44.3

April Total Precipitation/Normal: 3.52"/1.85"

Result: Below average temperatures,

Above average precipitation

May 6, 1965-Thursday:

HEADLINE-"Tornadoes Touchdown Briefly in State". Dozens of tornadoes swirled across Minnesota Wednesday. Funnels were sighted along a line extending from North Dakota southeastward across Minnesota. Spawned by a low pressure center in North & South Dakota, most of the tornadoes appeared north, south, and west of the Twin Cities. Damage was reported in Harmony, Parkers Prairie, Fargo, Detroit Lakes, Brainerd, Alexandria, Sauk Centre, Rochester, and Winona areas.The Weather Bureau predicts more storm activity in the Upper Midwest. Showers, thunderstorms, and gusting winds are forecast for large parts of the region. A deep low pressure area is moving northward across the Upper Midwest.Actual high/low: 78/57 with 1.11" of precipitation

May 7, 1965-Friday:

HEADLINE-"Tornadoes Roar Through Northeast Suburbs, Tonka Area Hard Hit". Tornadoes roared like a gang of drunken devils for six hours through the Minneapolis area Thursday night, killing at least 10, injuring hundreds, exploding homes and tossing trucks and cars around like toys. Warm, moist air flowing into Minnesota from the Gulf Coast and clashing with cooler air from the north produced Thursday's disastrous tornadoes. A cold front is located in the SE part of the state bringing fair to partly cloudy skies. Forecast highs for Saturday from the mid-60's to low-70's.Actual high/low: 77/58 with no precipitation

Synopsis of May 2nd-May 7th, 1965

Sunday, May 2, 1965-
Unseasonable Warm Temperatures

Monday, May 3, 1965-
Hailstorms Plague Area

Tuesday, May 4, 1965-
Cool Down-Rain

Wednesday, May 5, 1965-
Dozens Of Tornadoes Hit Outstate Minnesota

Thursday, May 6, 1965-
Tornadoes Devastate Twin Cities

Friday, May 7, 1965-
Cool Down-Clearing Skies

Note:

Weather maps (1965) show weather

conditions/predictions a day earlier.

State's Most Destructive Tornadoes prior to May 6, 1965

July 15, 1881

New Ulm, 4:45 p.m.

6 lives lost

Property damage $400,000.

August 21, 1883

Rochester, 6:36 p.m.

31 lives lost

Property damage $200,000.

The Mayo brothers, from Le Sueur, come to aid in the aftermath eventually starting the famous Mayo Clinic.

April 14, 1886

Sauk Rapids, 4 p.m.

74 lives lost, 136 injured,

Property damage $500,000.

This monster, long track tornado when crossing the Mississippi was reported to suck dry to river bottom and continued through Rice ending up near Buckman.

August 20, 1904

Twin Cities, 8 p.m.

14 lives lost,

Property damage $1,500,000.

August 21, 1918

Tyler, 9:20 p.m.

36 lives lost,

Property damage $1,000,000.

Most of the business section of Tyler completely demolished.

June 22, 1919

Fergus Falls, 4:45 p.m.

59 lives lost

Property damage $3,500,000.

This twin tornado cluster devastated downtown Fergus leaving so much debris on Lake Alice that "one could walk across the lake without getting your feet wet".

June 18, 1939

Anoka/Champlin, 3:10 p.m.

9 lives lost, 222 injured, property damage $1,200,000, greatest damage occurred in Anoka where there was a path of destruction three blocks wide affecting 240 families.

July 20, 1951

Hennepin County, 9:26 p.m.

5 lives lost

Property damage $6,000,000,

Greatest damage occurred at the Wold-Chamberlain Field where 63 planes were destroyed and 37 others damaged.

The Fatalities of the May 6th, 1965 Tornadoes

Helene Hawley, 4 months, 1075 N. Circle Drive, Fridley

Walter Achterkirch, 67, 2001 County Road H, Mounds View

Lavila Jean Abraham, 32, 2255 Lois Drive, Mounds View

Lori Ann Abraham, 4, 2255 Lois Drive, Mounds View

Lisa Abraham, 1 month, 2255 Lois Drive, Mounds View

Robert E. Clarke, 26, Route 5, Anoka

Clarence Paulsen, 65, Route 3, Mound

Wilma Paulsen, 60, Route 3, Mound

Alma Grinde, 841 Valentine Lane, Blaine

Mrs. John Iverson, 80, Island Park, Minnetonka

Gregory Magsam, 4, 2249 Lois Drive, Mounds View

Raymond Perbix, 56, Norwood

Annie Demery, 64, 999 Pandora Drive, Fridley

Baby Fossum, still born 5/7/65, 7310 Concerto Curve, Fridley

Differences Between Watches and Warnings

This discussion is cause for concern. Many citizens are confused by the words Watch vs. Warning and many times do not take heed of the warnings when they are issued. False alarms and radar-indicated tornadoes only add to the confusion. Following is an explanation of each situation

▽ **Severe Thunderstorm Watch:**

Affected area issued by the Storm Prediction Center. Conditions are favorable for the development of severe thunderstorms. By definition, severe thunderstorms produce hail ?" in diameter and/or wind gusts in excess of 58 mph.

▽ **Severe Thunderstorm Warning:**

Warnings are issued by the local National Weather Service offices. The warning means that a severe thunderstorm is occurring or imminent. The affected area is issued by counties or portions of counties. These warnings typically last for a period of 30-60 minutes. When issued for your area, seek shelter immediately. The local municipalities will sound the storm sirens.

▽ **Tornado Watch:**

Affected area issued by the Storm Prediction Center. Conditions are favorable for the development of severe thunderstorms and multiple tornadoes. Persons in affected areas should be vigilant in watching for threatening weather.

▽ **Tornado Warning:**

Warnings are issued by the local National Weather Service offices. This warning means a tornado has been sighted by trained observers or radar indicated. The affected area is issued by counties or portions of counties. When issued for your area, seek shelter immediately. The local municipalities will sound the storm sirens.

What to do in the event of Severe Weather

Tornadoes are extremely complex wind events that cause damage ranging from minimal or minor to absolute devastation. Seeking shelter immediately when a warning is issued is the best defense to avoiding injury or loss of life. Sirens may or may not sound due to the proximity of the storm, sometimes tornadoes have rendered the sirens silent due to the damage inflicted. Listed in the following section are recommendations for a variety of situations. Home, school, work, and type of structure all determine different strategies for safety. The number one rule to remember is to "get low!"

The home provides adequate protection from tornadoes and severe thunderstorms. If there is a basement present, get to that area and protect yourself and others by going under

sturdy furniture or under the stairwell. Debris is most likely to fall downward and by seeking shelter under a protective shell, the risk of being injured by falling debris is minimal. If there is no basement, get to the lowest level of the home and seek shelter in an interior, windowless room, preferably a bathroom. In violent tornadoes, the interior bathroom is the only structure left standing despite the massive destruction of the peripheral rooms. If you live in a trailer home, get out and seek shelter in the designated complex storm shelter. If there is little time, get into a ditch or low lying area and cover your head with your arms. Trailer homes offer little protection from tornadoes or severe thunderstorms. The mortality rate for trailer parks is higher than any other structure. Apartment buildings offer the greatest protection in the lowest level away from windows in an interior hallway or bathroom.

Schools today are much more crowded and with the advent of "portables", the danger to students is higher than in the past. Most schools have tornado practice drills which help guide students and staff to an appropriate safe area. "Portables" offer the least amount of protection and similar to trailer homes, people should get out of these structures and seek shelter in a more reinforced building. If little time, "get low!" Seeking shelter in the second floor of the school should be avoided as well. Violent tornadoes have been known to rip off the roofing and supporting structure to leave occupants exposed to flying debris. The first floor interior/windowless hallways offer the greatest protection. Any school areas with roof spans of 40 feet or more should be avoided as well. This type of roofing structure offers little support and is often lifted off and ripped away by violent tornadoes. Auditoriums, lunchrooms, and field houses are prime examples to avoid seeking shelter. Malls have similar roof design with wide, open spaces. These places should be avoided and seek the designated storm shelter area. Bright, attention-getting signs should inform shoppers of the availability of these shelters.

At work, seek shelter in the lowest possible level avoiding windows. Many places of employment offer a safe area to go to in case of severe weather. Basements, interior/windowless bathrooms, reinforced storm shelters all offer adequate protection. It is recommended that periodic practice drills be performed to inform new employees and re-familiarize existing staff as to seeking shelter during tornado or severe weather outbreaks. Research has shown that persons seeking shelter during these outbreaks have a tremendous survival rate over uninformed persons.

If you are in your car, "get out!" Seek shelter in a low area or ditch. Never try to outrun a tornado! Tornadoes have been "clocked" at more than 70 miles per hour! You run the risk of being involved in an accident trying to flee a tornado or simply being hit by the extreme winds of the tornado! In Andover, KS, a tornado traveling down the interstate caught hold of a minivan and flipped it around the funnel base like a "tumbleweed."

A mile wide F5 tornado struck Moore, Oklahoma in 1999 where 45 people tragically lost their lives. Out of that population, no school age children between the ages of 5-18 lost their lives. This statistical anomaly attests to the concept of weather preparedness. These young people sought the appropriate shelter and were protected from this violent tornado. This is a prime example of informed people acting on their knowledge to save their lives.

To review

- The most important aspect of severe weather safety is "get low!"
- Seek shelter away from windows and wide, open roof spans.
- Cover your head with your arms and crouch to the lowest possible profile.

One of the leading causes of death in tornadoes is from flying debris

GET LOW!

Please copy and post.

Practice Tornado safety.

Official Mounds View List of Condemned, Destroyed, and Seriously Damaged Homes - May 17, 1965

2296 Lois Drive	Armstrong, S.	Destroyed
2292 Lois Drive	Ingram, D.E.	Destroyed
2286 Lois Drive	Lubbe, A.	Destroyed
2278 Lois Drive	Johnson, E.D.	Destroyed
2272 Lois Drive	Orton, L.R.	Destroyed
2264 Lois Drive	Meyer, E.F.	Destroyed
2256 Lois Drive	Gifford, R.G.	Destroyed
2250 Lois Drive	Czech, R.V.	Destroyed
2242 Lois Drive	Sachs, O.L.	Destroyed
2234 Lois Drive	Lohse, N.H.	Destroyed
2226 Lois Drive	Snyder, T.	Destroyed
2208 Lois Drive	Pittman, J.W.	Destroyed

2204 Lois Drive	Anderson, B.L.	Destroyed
2200 Lois Drive	Anderson, W.R.	Destroyed
2203 Lois Drive	Brown, C.	Destroyed
2209 Lois Drive	Pasiowitz, J.F.	Destroyed
2215 Lois Drive	Angell, C.	Destroyed
2249 Lois Drive	Magsam, M.R.	Destroyed
2255 Lois Drive	Abraham, C.F.	Destroyed
2261 Lois Drive	Erickson, D.G.	Destroyed
2267 Lois Drive	Young, R.J.	Destroyed
2273 Lois Drive	Germain, B.J.	Destroyed
2279 Lois Drive	Spriggle, S.C.	Destroyed
2285 Lois Drive	Madsen, D.W.	Destroyed
2291 Lois Drive	Burg, D.	Destroyed
2297 Lois Drive	Rossbach, D.J.	Destroyed
2303 Lois Drive	Buck, T.	Destroyed
2291 Hillview Rd.	Lind, L.V.	Destroyed
2299 Hillview Rd.	Sherman, T.G.	Destroyed
2309 Hillview Rd.	Lambert, M.C.	Destroyed
2317 Hillview Rd.	Koenker, M.	Destroyed
2327 Hillview Rd.	Anderson, R.P.	Destroyed

2335 Hillview Rd.	Skjod, L.W.	Destroyed
2343 Hillview Rd.	Carufel, D.J.	Destroyed
2310 Hillview Rd.	Whiteneck, R.	Destroyed
2316 Hillview Rd.	Baty, F.	Destroyed
2324 Hillview Rd.	Christensen, E.	Destroyed
2334 Hillview Rd.	Beckman, R.J.	Destroyed
2342 Hillview Rd.	Aronson, D.J.	Destroyed
2350 Hillview Rd.	Sullivan, R.	Destroyed
2340 Knoll Drive	Fodstad, D.	Condemned
2334 Knoll Drive	Klein, C.	Damage
2337 Knoll Drive	Braley, J.	Damage
2333 Knoll Drive	Thornton, D.	Condemned
2328 Knoll Drive	Johnson, K.A.	Damage
2324 Knoll Drive	Beach, G.D.	Repair
2325 Knoll Drive	Cheney, A.W.	Damage
2319 Knoll Drive	Koelndorfer, M.A.	Damage
2318 Knoll Drive	Hamilton, D.	Destroyed
2313 Knoll Drive	Chapman, S.P.	Damage
2310 Knoll Drive	Guzzo, R.	Destroyed
2302 Knoll Drive	Johnson, E.V.	Destroyed

2307 Knoll Drive	Johnson, R.L.	Damage
2296 Knoll Drive	Wendorf, W.	Destroyed
2299 Knoll Drive	Pickar, J.E.	Damage
2288 Knoll Drive	Carlson, D.M.	Damage
2291 Knoll Drive	Jurgensen, M.	Damage
2283 Knoll Drive	Wilke, C.	Damage
2280 Knoll Drive	Juve, R.D.	Condemned
2277 Knoll Drive	Peterson, R.A.	Damage
2274 Knoll Drive	Overgaauw, J.	Destroyed
2271 Knoll Drive	Ames, J.F.	Destroyed
2268 Knoll Drive	Sherff, E.R. Jr.	Destroyed
2265 Knoll Drive	Kruetter, R.D.	Condemned
2262 Knoll Drive	Lodien, F.	Destroyed
2259 Knoll Drive	Thompson, D.D.	Condemned
2256 Knoll Drive	Meyer, L.	Destroyed
2253 Knoll Drive	Scheid, A.A.	Destroyed
2250 Knoll Drive	Blaine, A.E.	Destroyed
2249 Knoll Drive	Fitting, L.S.	Destroyed
2245 Knoll Drive	Peterson, W.L.	Destroyed
2241 Knoll Drive	Tabery, J.E.	Destroyed

2237 Knoll Drive	Freestone, L.M.	Destroyed
2233 Knoll Drive	Bohler, L.C.	Destroyed

(Fairchild Avenue is now Silver Lake Road)

7440 Fairchild Av.	Waschek	Damage
7408 Fairchild Av.	Fagerstrom	Damage
7350 Fairchild Av.	Varns, J.F.	Destroyed
7360 Fairchild Av.	Olund, I.	Destroyed
7338 Fairchild Av.	Johnson, A.	Damage
7404 Fairchild Av.	Pederson, L.I.	Destroyed
7305 Knollwood Dr.	Tom Thumb	Condemned
1884 County Rd. H2	Staurseth, E.	Condemned
1891 County Rd. H2	Achterkirch, W.O.	Destroyed
1893 County Rd. H2	Pueringer, I.	Damage
1866 County Rd. H2	Imm, G.	Condemned
1865 County Rd. H2	Miller, R.C.	Damaged
1707 Hillview Rd.	Ramsey Co. Library	Condemned
2525 County Rd. I	Gustafson, E.E.	Condemned
2545 County Rd. I	Bakken, R.	Destroyed
2549 County Rd. I	Gustafson, E.E.	Destroyed
2520 County Rd. I	Sunrise Church	Condemned

2520 County Rd. I	Parsonage	Damage
County Rd. I & Woodlawn	8-plex apartment	Condemned
7701 Woodlawn Dr.	Hembre, H.	Damage
7710 Woodlawn Dr.	Staples, J.	Destroyed
7768 Woodlawn Dr.	Parker, J.L.	Destroyed
Quincy Rd.	4-plex apartment	Destroyed
Quincy Rd.	4-plex apartment	Condemned
2259 Oakwood Dr.	Ctvrtnik, J.	Damage
2287 Terrace Rd.	Monson, E.	Damage

Final Damage Amounts From Twin Cities Tornadoes

Anoka County	$36 million dollars
Fridley	$23 million
Spring Lake Park	$12 million
Blaine	$ 1 million
Hennepin County	$8 million dollars
(no breakdown given)	
Ramsey County	$3.25 million dollars
(Mounds View/Shoreview)	
Carver County	$4.25 million dollars
Chanhassen	$1,320,000
Young America, Waconia, Benton, and Dahlgren	$2,930,000

Total Damage

in 1965 dollars	$51.5 million dollars

Total Damage

in 2007 dollars	$ 1.2 billion dollars
in 2010 dollars	$ 3.5 billion dollars

311 homes destroyed

609 major damage to homes

551 minor damage to homes

61 businesses destroyed

277 trailer homes destroyed

95 farm buildings destroyed

In terms of human cost and suffering, the results are absolutely devastating: at least 14 people lost their lives and 683 injured. Thousands more still suffer from the ramifications of those storms in terms of anxiety and fear of storms. This swarm of tornadoes was one of the costliest storms ever to hit the United States.

Interesting Relevant Websites

Outstanding Historical Minnesota Flood Information Site. First, you need to Google or Yahoo search "1965 Mississippi Floods". When you get to that site, cursor down to this link listed below. http://www.murphylibrary.uslax.edu/digital/lacrosse/missfloodplain/text.html

- Excellent Local Radar Site
  www.weather.com
- Historical White Bear Lake Ice Out Information
  www.wblcd.org/WhiteBear-LakeIceOutDates.html
- Official NOAA Tornado Preparedness Site
  www.crh.noaa.gov
- Excellent Tornado Information Site
  http://itlnet/tornado

- Fantastic Minnesota Historical Society Site www.mnhs.org/localhis-tory/mho/chsclo.html
- Great Tornado Information Site http://www.tornadoproject.com.html
- Interesting Minnesota Cities Information http://www.citypopulation.de/USA-Minnesota.html
- Informational Official National Weather Service Site www.nws.noaa.com
- Information-packed 1965 Twin Cities Tornadoes Site http://www.crh.noaa.gov/mpx/HistoricalEvents/1965May06/index.php#radar
- Wonderful Collection of Historic Tornado Photos Albums by National Severe Storms Laboratory (NSSL) www.photolib.noaa.gov/nssl/#album
- Fantastic Civil Defense Siren/Historical Sounds Site www.civildefensemuse-um.com/sirens.html

Absolutely Breathtaking

Actual WCCO Radio recordings from May 6th, 1965 (parts 1 & 2) http://www.radiotapes.com/user/WCCO%20Tornado%205-6-1965%20-%20Part%201.mp3

AND

http://www.radiotapes.com/user/WCCO%20Tornado%205-6-1965%20-%20Part%202.mp3

About The Author

Allen W. Taylor is a mathematics educator and author who has been interested in chasing and researching tornadoes since May 6th, 1965.

He lives with his wife and children in the Twin Cities area of Minnesota.

Mr. Taylor is available for community presentations and speaking.

For quantities or wholesale, please call 763-422-3878